B SOMORJAI, G. A.
Somorjai, Gabor A.
An American Scientistt
autobiography of Gab

An American SCIENTIST

Gabor Somorjai

An.
American
SCIENTIST

The Autobiography of
GABOR A. SOMORJAI
with Mitch Jacoby

Gabor Somorjai

 ARCHWAY

Archway Publishing books may be ordered through booksellers or by contacting:

Archway Publishing
1663 Liberty Drive
Bloomington, IN 47403
www.archwaypublishing.com
1-(888)-242-5904

ISBN: 978-1-4808-0146-2 (sc)
ISBN: 978-1-4808-0147-9 (hc)
ISBN: 978-1-4808-0145-5 (e)

Printed in the United States of America

Archway Publishing rev. date: 09/13/2013

The spectacular rise of science and the technologies it spawned in the last fifty years of the twentieth century and beyond—and their pronounced influence on the health, wealth, and quality of life in the U.S. and the world, as told through the life story of a Hungarian-born scientist

CONTENTS

PREFACE

Riding The Waves

Near the golden sands of Kaanapali Beach on Maui, a little boy takes on the waves. Sometimes he rides atop the crests, sometimes he dives under them, and only rarely do the waves win and dump him on the beach. He loves this game and plays it well because he has the confidence to quickly decide which way to go.

The little boy is my grandson, Benjamin. His beachfront competition with the deep blue ocean brings to mind my seven decades' of riding life's waves. Sometimes the ocean was calm and sometimes it brought gently rolling waves. But other times the seas were rough I had to plunge headfirst into choppy waters—and I wasn't always sure if I would come up for air. But I always did.

I don't really know why I have always managed to come back up. It's hard to say. Perhaps it's just instinct—the will to survive. But maybe it's because I am always looking forward—looking to the future instead of dwelling in the past. Like Benjamin, I am always looking in the direction of the next wave.

THE PATERNOSTER

As a three year old, I thought the *paternoster* was magical.

This marvelous early-style elevator went up and down without stopping—shuttling people, one at a time, between the floors of the Hungarian Credit Bank, in Budapest, where my father worked when I was a little boy. The moving metal cubicle didn't come to a quiet stop and wait for passengers to load and unload leisurely. Rather, as it approached our floor, we had to get ready to step in—just in time—and quickly step out when we reached our destination.

The trips to the imposing bank building and the rides on its magical paternoster were exciting events in the early part of my childhood. It was a happy time and a wonderful place for a little boy to grow up —and I was fortunate to be there in that place at that time with my family—though little kids don't usually think about their good fortunes. But good fortune is indeed what I had and it's something that stayed with me even when life became stormy, which it did more than a few times—like during the mayhem that turned Hungary and its people upside down some 65 years ago.

I had the good fortune to come from a strong family and from a country that produced some of the greatest scientific minds of the first half of the 20th century. And when all hell broke loose there in the 1950s, I was lucky enough to have escaped—and to have done so together with Judith Kaldor, with whom I have been blessed to raise a beautiful family and share more than 55 years of marriage.

As a post World War II refugee who fled to the United States, I was again fortunate to have made my way to the west coast, where I was accepted as a graduate student at the University of California, Berkeley, and later became a chemistry professor—a position I have now held for nearly five decades.

Without a doubt I ended up in the right place and at the right time. As the United States became the science and technology engine of the world, Berkeley, home to a top-flight research university and Lawrence Berkeley National Laboratory, served as one of its main powerhouses of scientific innovation and discovery. And I was positioned right there in the middle of it.

My research career, based on an area of chemistry known as surface science, coincided directly with the rise of this country's prowess in high-tech industries, such as microelectronics, computer disk drives, and other revolutionary and game-changing fields. My career also overlapped with the onset of the energy crisis of the 1970s, which triggered a wave of research in a surface-chemistry specialty area known as catalysis. That field, of which I have long been an active and proud proponent, took upon itself the challenge of discovering the catalytic materials and methods needed to produce synthetic fuels and to make the most of natural resources while using as little energy as possible. That's also the field of science that developed the automobile catalytic converter and other devices and techniques for protecting the environment and cleaning the air, water, and soil.

By building a research program in these thriving and exciting areas of science that have helped improve this country's—and for that matter, the world's—health, well-being, and standard of living, I had the good fortune to attract top students and research associates from all over the world to my Berkeley laboratory.

Like many other American scientists, my scientific genealogy or family tree originates in Western Europe, specifically Italy and Germany. In the usual ways of scientific continuity, mentors pass along to their students thought processes, concepts, and approaches to thinking about nature. For generations, scientists have learned from their predecessors, developed new ideas as they unfolded in their day, and passed along new knowledge to the next generation. I have been lucky to be part of the time-honored tradition of scientific pedagogy.

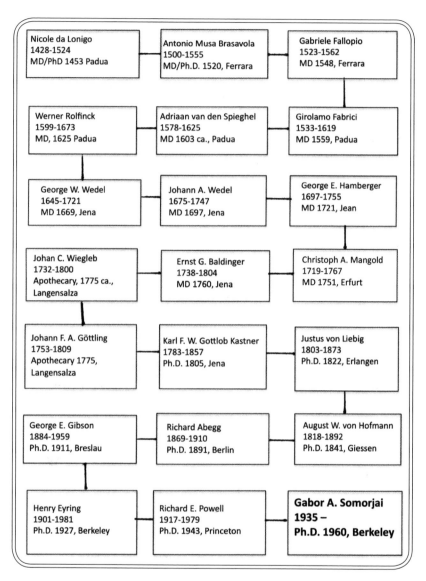

*The roots of my scientific genealogy stretch from
Italy and Germany to the United States.*

After a long and fruitful career that has been highlighted by international recognition and honor, I deeply believe that my most important accomplishment by far, and the one of which I am most proud, is this multinational, multicultural, and multigenerational web of some 400 interconnected scientists who make up the extended Somorjai research group.

A number of world famous Hungarian scientists came to the United States in the early part of the 20ᵗʰ century and helped strengthen this country scientifically, technologically, and militarily. Among the most famous are Edward Teller, Eugene Wigner, Theodore von Karman, John von Neuman, and Leó Szilárd. I'm proud to hail from the same country of origin as these seminal thinkers and like them, to have pursued a career of scientific discovery. But this group of scientific superstars predated me by a few decades.

My life history is centered on the next fifty years, roughly the second half of the 20ᵗʰ century, when a second generation of Hungarian-born American scientists made its mark on U.S. science and technical know-how. Like all immigrants, I faced cultural and professional challenges, the outcomes of which often determined success or failure. Here again I was fortunate to have made some good initial choices that gave me the self-confidence to take risks, be creative, and recognize rewarding opportunities.

The latter generation of Hungarian scientists, the one in which I include myself, faced political and social challenges that were very different than the challenges faced by the earlier generation. Yet like the scientists who preceded them, my contemporaries also thrived while making important and lasting contributions to the U.S.

I'm not going to tell you that I singlehandedly sent man to the moon or invented personal computers or the Internet, for that matter. I'm not even going to tell you that things were always easy. They weren't. Like the rides on the old-time paternoster, there were times in my career that I was clearly moving up and other times that I seemed to be coming down. But I *will* tell you—with pride—that the discoveries my students and I made together have in some ways helped advance science and technology and have helped make progress in addressing important societal needs. I will also tell you that the blessings of good health and my family's love have made this long trip possible.

But I'm getting ahead of myself now. Let me start from the beginning and tell you a little about my family and what life was like for a young boy growing up in war-time Hungary.

CHAPTER 1

A Family Tree with Deep Roots

I remember pretty furniture and broad bookshelves. I remember Sunday dinners with my grandparents and the warmth of freshly baked bread. I remember vacationing in the Hungarian countryside and the joy of showing off that I had learned to cuss. I remember black boots and soldiers' uniforms, yellow stars and marches along the Danube. And I remember feelings of anxiety, the curse of helplessness, and the unforgettable smell of death.

IN THE LATE 1800s, my mother's father headed west from what is now Romania to seek his fortune in Budapest. He came from a family of Sephardic—literally "Spanish" Jews, a group that traces its roots back to late 15th-century Spain, to the time and place of the Spanish Inquisition. At that time, Spain's Catholic monarchs,

Ferdinand and Isabella, sought to ensure the fidelity of recent converts to Catholicism by forcibly expelling Jews and other non-Catholics.

The expulsion from Spain scattered Sephardic Jewry. Some groups chose to remain in Christian Europe while others headed east to the Ottoman (Turkish) Empire or to Morocco, lands that were predominantly Islamic. In those times, Jews often faired better in Islamic countries than in Christian ones, and so my grandfather's ancestors came to settle in Ottoman lands.

As history records, the Ottoman Empire expanded mightily during the 1400s and 1500s, growing in size, power, and influence. That expansion eventually absorbed much of Hungary following the Ottoman defeat of the Hungarian army at Mohács in 1526 and the capture a short time later of Buda, which lies on the western bank of the Danube River in modern Budapest.

For a century and a half, Hungary remained largely under Ottoman control. But the tide began turning by the early 1700s, and over the next 200 or so years, the Ottoman Empire gradually declined. The shift in power led to sweeping changes throughout central Europe and new opportunities for prosperity. My grandfather looked to the Austro-Hungarian Empire and saw revitalization, a thriving economy, and the chance to make a better life for his family.

And so it was that around 1880 Moric Ormos, my maternal grandfather, moved to Budapest from Nagyvárad, which is the Hungarian name of the Romanian city Oradea. He soon opened a shoe store there and later a second store. In time, his brother in law, my Karoly Bacsi, or Uncle Charles, went into shoe manufacturing, having learned that trade in Prague and London. Overall, things went well for my grandfather. You couldn't exactly call him wealthy, but he managed just fine. He established himself solidly in Budapest's mercantile class and in its Sephardic community and did well enough to buy a two-story apartment building, where I lived from my birth in 1935 until my escape from Hungary in 1956.

My paternal grandfather, Herman Steiner, came from Austria. Family records, which date back to 1812, show that my father's side of the family traces its roots to Grossensersdorf, a small Austrian town known today as Gross-Enzersdorf, just east of Vienna.

My grandfather was an engineer who built roads for the Austro-Hungarian Empire. As a civil servant, he traveled often and ended up working in Hungary, where he met and fell in love with a Jewish Hungarian girl, my grandmother, Fanni Edelstein. Perhaps to please her family, or maybe as a precondition to taking her hand in marriage, my grandfather, converted from Catholicism to Judaism and married my grandmother in 1891.

My father was born in 1902—and just a few years later, his mother died in childbirth. Her passing left my grandfather with six children to raise singlehandedly. He never remarried. To my grandfather's credit, my father and all of his siblings graduated high school, which was by no means the norm in those days. I don't remember much about my father's father. But I know he came from a distinguished Austrian family and never learned to speak Hungarian. He died in 1939 when I was four years old.

My father was a genius with numbers. He had a real gift—the kind of talent that no doubt would have made him an outstanding economist. But it was not to be. 1920 was the year my father graduated high school and the year he applied to university. It was also the year the nationalist government of Miklós Horty came to power in Hungary and the year Hungary imposed *Numerus Clausus*.

These "closed number" laws restricted the number of Jews admitted for higher education to about 5% of a university's incoming class. Although the justification for this religious (or "racial") quota system was portrayed innocently enough as a means of balancing Hungary's university and professional populations with the demographics of its national population, its real purpose seems clear.

Historians note that the 1920 policy is generally regarded as 20th-century Europe's first Anti-Jewish Act, not to mention the policy that drove Edward Teller and scores of other Jewish Hungarian scientists and intellectuals abroad. As history would unfold, this discriminatory measure would later be seen as but a drop in the bucket compared with the overtly anti-Jewish policies enacted when Hitler took the reigns of power in 1933. In hindsight, the 1920 events were a cold foreboding of the far-worse fate that would soon befall Hungarian Jewry.

So with little choice in the matter, my father opted for plan B—attending the more pedestrian business school instead of the more prestigious university. Nonetheless, he did well academically and was trained to pursue accounting or banking. Upon graduating, he landed a job at the Hungarian Credit Bank, a well-respected financial institution in its day.

His talents were recognized and he was promoted quickly. In fact, he rose through the ranks so rapidly, that by 1930, at the age of just 28 and just a year after he married my mother, he became one of the bank's junior directors.

My early childhood memories are rather spotty but some scenes still play in my mind vividly. I remember, for example, at about age three or four, the excitement of coming to pick up my father from work at the bank and riding in the amazing paternoster—the non-stopping elevator. Coming with my mother and sister, who is two and a half years older than me, to visit my father at work was exciting in its own right. But for a little kid, riding the paternoster was just thrilling! (As it happens, I visited Budapest again recently and stopped in at the Credit Bank. Amazingly, I found the paternoster still in operation after all these decades.)

In the mornings, after my father left for work, my sister and I would go for a walk with our nanny—our *tante*—to the nearby museum garden, where we played with her and with other children. After our daily regimen of exercise and outdoor playtime, we returned home for various kinds of lessons. I practiced accordion and learned to step dance—and even put on the occasional performance or participated in a recital for my admiring family.

In those days, we used to spend several weeks during the summer in the town of Siófok along the southern shore of Hungary's Lake Balaton. We stayed in a villa with my mother and nanny. My father would take the train and join us on the weekends. With its sandy beaches and small-town appeal, Siófok, which is only about 60 miles outside of Budapest, was a popular summer vacation destination for city folks. It still is.

One summer, instead of hanging out on the beaches of Lake Balaton, we spent a few weeks vacationing on The Puszta. This vast region of grasslands, open plains, and wetlands, which today is the site of national parks, has been a prime location for ranching and cattle grazing for centuries.

We stayed at the estate of one of our parents' friends—a spacious place with horses, pigs, and cattle. It would make for a pretty good story to tell you that out there in (Magyar) cowboy country, as a little kid, I learned to rope a steer. But it wouldn't be true. I did pick up another memorable skill, however. That summer out on The Puszta, I learned to swear.

Every language has its collection of cuss words. But the Hungarian collection is especially rich and colorful. If you really know how to swear like a sailor in Hungarian, you could keep at it for quite a while without repeating yourself, owing to the richness of the language. In any case, during our stay, I befriended some of the boys in the local village, hung out and played with them, and learned a whole bunch of naughty words.

The reason it was so memorable is that upon returning home, my grandfather, who lived one floor below us, somehow caught wind of my newfound vocabulary. He was so proud of his little grandson with the potty mouth, that he paid me a few coins to show off all the bad words I had learned that summer.

Living so close to my grandparents—my mother's parents—was a blessing. I remember running around the little balcony—the walkway that encircled the inner courtyard of our building and calling out to my grandmother whenever I smelled fresh baking. "Gramma (nagyanya) can I come and have a piece?" Of course you can, she would answer me. When it comes to feeding their grandchildren, grandmothers seldom say "no."

Freshly baked cake and warm bread weren't the only things I enjoyed at my grandparents' home. We regularly gathered there on Sundays for dinner. These family gatherings—together with two of my uncles—were some of the best times of my childhood.

After dinner, the women, together with my sister, would go off to the living room to have tea or I don't know?—to do whatever it is that the ladies did. But *I* got to stay behind with the men and do manly things like "help" them play cards. A couple of times my enthusiasm got the better of me and I guess I "helped" a bit too much—unintentionally announcing to the room who held which cards. As you might guess, that didn't go over well at all.

*My extended family at the dinner table on the first floor of
our house in Budapest, following my parents wedding in 1929.
Sitting at the head of the table is my paternal grandfather.
Behind him, stand my parents the newlyweds. On each side of
them are my maternal grandparents and great grandparents.*

Like my grandparents' apartment, ours also had beautiful
bookshelves encased with glass doors that displayed handsome
clothbound books. As a small child, I admired their pretty covers
and, later on, when I was forced to spend a lot of time indoors, I
pored over their contents and learned a lot about the world, without
actually setting foot outside of Hungary.

We lived in relative comfort in one of the second floor
apartments of my grandfather's two-story building at 29 Maria
Street in Budapest's District no. 8. We had a live-in maid, a girl from
the countryside, who helped with the house work and took care
of my sister and me—and there was a "housemeister," a handyman

of sorts who, together with his wife, lived on the ground floor and looked after the building maintenance and the supply of wood and coal. Anyone reading this story takes for granted the convenience of having hot water for bathing. But when I was a youngster, the maid—or later, after we were forced to let the maid go, my mother had to boil water on the stove and haul it to the bathroom to fill the tub so we could take a hot bath.

Our apartment was spacious—not huge by any stretch of the imagination, yet it had some rooms in which I spent almost no time at all. For example, adjacent to the room in which we generally ate our meals (which, by the way, doubled each night as a bedroom for the children), was a pretty living room that my parents used when we had visitors or when they entertained guests. That room led into a formal dining room that we generally used only on special occasions such as holiday dinners that were attended by our extended family members. The room was beautifully decorated with ornate carved furniture—furniture that would later prove to be too heavy to enable my family to move from the apartment easily.

Overall, my earliest years were happy ones indeed. We lived in close proximity to extended family members we loved and with whom we enjoyed spending time. And we lived in a nice apartment in the handsome two-story house that certainly must have been a source of pride—a symbol of sorts to my grandfather and to our entire family that we were living among Budapest's upper middle class. But alas, those days were numbered. As fate would have it, our family's comfort and security would soon crumble.

Changes were afoot. Around 1939, as Hungary began aligning itself with Germany, the Credit Bank fired my father along with all the other Jewish employees. So he went into retail business, opening a shoe store like my grandfather. Although I know now that major events were unfolding in Europe at that time—most prominent among them, Germany had invaded Poland marking the start of the Second World War—little kids don't usually know about such things, and I was no different.

My parents, my sister Marietta, and I in the early 1940s.

The biggest changes in my life were that my mother wasn't always around because she had begun working regularly in the store with my father—and our *tante* left. As it turns out, my parents couldn't really afford to keep the nanny after my father lost his job, but they didn't have much choice in the matter. As a German woman, the nanny was not permitted to work for a Jewish family.

Then came significant changes. My father was drafted into the Hungarian army. At first, as an officer, he was dispatched along with Hungarian troops to retake a portion of Banat in southeastern Hungary. That region was part of Hungary before World War I. But as a result of the Treaty of Trianon in 1920, the land became a piece of Yugoslavia. (Today it's parceled between Romania, Serbia, and Hungary.) Early in World War II, Hungary was eager to reclaim the land—and they did so with help from Hitler. That acquisition bolstered Hungarian nationalism and cemented Hungary's relationship with Germany.

Not long thereafter, new anti-Semitic laws were enacted in Hungary. My father was stripped of his officer status and reduced to common laborer. As Germany moved to invade Russia, the Hungarian army joined up with German forces on the eastern front and my father's situation deteriorated rapidly. Jewish Hungarian troops were amassed in a work detail and sent off to the front lines to deactivate landmines—an assignment from which very few returned.

Jumping forward a bit in time, I'll tell you that somehow, my father managed to survive through the worsening conditions of 1943 and 1944. He eventually ended up incarcerated in the Nazi concentration camp complex in Mauthausen, Austria. Despite meticulous German record keeping, exact numbers for the inmate death toll are unknown due to Germany's deliberate efforts to destroy files and records shortly before abandoning the camp in the spring of 1945. Nonetheless, historians generally put the loss of life at greater than 200,000 people.

Even after American forces liberated that horrible place on May 5th 1945, inmates continued dropping like flies as their ravaged bodies finally succumbed to disease, starvation, and exhaustion. Of some 10,000 Hungarian men forced into the Jewish labor brigade, by the time the atrocities ended, maybe 50 of them were still alive. Perhaps due to tremendous willpower, or good fortune, or who knows what, by the middle of 1945, my father returned home. By then he was a mere skeleton, suffering from typhoid, and weighing less than 100 pounds.

Meanwhile, life in Budapest had turned sour while my father was gone. First we lost my grandfather. In the fall of 1943, he stepped out of his shoe store to get a haircut across the street and collapsed—dying suddenly of a massive heart attack. Who could have realized at the time how lucky he was to have been spared the heartache and misery that was to befall all of us?

Then on March 19, 1944, German troops invaded Hungary after Admiral Horthy, the Hungarian dictator, tried unsuccessfully to break his alliance with Germany. Along with the German troops came Adolf Eichmann, the infamous senior Nazi SS man who would come to be known as the "architect of the Holocaust."

Endowed with outstanding managerial and organizational talents, Eichmann had by that time already choreographed the complex mass deportation of Jews to ghettos and extermination camps across German-occupied Eastern Europe, in the Greek port city of Salonica, and elsewhere.

Upon invading Hungary, Hitler was deeply concerned about Russia's military position. The Red Army was advancing steadily on Budapest and the opportunity to rid Europe of its Jews was quickly vanishing. So Hitler sent his best man, Eichmann, to deal once and for all with "the Jewish question" in Hungary.

Measured by the twisted yardstick of Nazi Germany, he would turn out to be wildly successful. According to Eichmann's own records, in a period of just three weeks he arranged for the transport of some 437,000 Hungarian Jews by train to their deaths in the Nazi gas chambers.

Just days after Germany swiftly moved to occupy Budapest, a Gestapo car pulled up in front of our house. Two officers had come to take away my grandfather. They did not know he was already dead. They had a list of names of Jewish community leaders, well-to-do businessman, and other prominent Jews—and my grandfather's name was on that list. When the Gestapo officers left, it was clear to my family that Hungary as we knew it—especially life for Hungarian Jews had changed forever.

Within a month, whenever we stepped outdoors, we were required to wear yellow Stars of David identifying ourselves as Jews.

And at that point, a number of houses in our neighborhood, ours among them, were also marked with the yellow stars to make it clear to passersby which houses were owned by Jews.

The rhythm of daily life changed quickly. I was no longer going to school because Jews weren't allowed there. I also wasn't playing much with friends, because playing outside at that time was ill-advised. If I look back now at that time and search for the silver lining hiding behind the clouds, I would say that being cooped up indoors led me to discover my family's extensive book collection and the literary treasures they held. The experience also taught me to play chess.

An older gentleman, a doctor who was a distant relative of ours, lived on the first floor of our building. He was a kind and gentle person—an excellent chess player and a patient teacher. I spent my mornings with him, playing chess day after day honing skills that later in life would be used to devise scientific problem-solving strategies.

It was also then that my grandmother was diagnosed with leukemia and hospitalized. I suspect that in addition to her physical ailments, she was heartbroken by the loss of her husband, my grandfather. It could only have made matters worse, that Jewish shops were confiscated at that time. With their hands tied behind their backs, my mother and grandmother tried to entrust some of their most loyal employees with the responsibility of managing each of the stores. They had hoped that by scheming to make it appear that the stores were owned by non-Jews, they would be able to maintain some control of the businesses. It didn't work. These so-called loyal employees took over the businesses for themselves and promptly denied all knowledge of the rightful owners.

By August 1944, most of Budapest's Jews were ordered to move into a ghetto—an area measuring a few blocks square in the Pest half of the city on the eastern bank of the Danube River. It may seem strange to look upon a forced move to the ghetto as a stroke of good luck but that's exactly what it was. We would later learn that while some of Budapest's Jews were surviving in the ghetto, the rest of country's Jews, meaning nearly all of those who lived outside of Budapest, including most of my extended family, were rounded up, processed quickly, and shipped off to die at Auschwitz.

Meanwhile, for some of Budapest's lucky Jews, salvation and protection came in unexpected ways. Through the compassion of a handful of humanitarians, such as Swedish diplomat Raoul Wallenberg, Germans were persuaded in one way or another to leave a sizable number of Jews under Swedish protection. By bartering, negotiating, and trading Swedish products, Wallenberg's delegation managed to issue Swedish passports to thousands of Hungarian Jews. These people, ultimately about 20,000 in all, were granted a type of diplomatic immunity and housed in buildings established as Swedish territory and thereby protected.

Through persuasion and negotiation, Swedish diplomat Raoul Wallenberg secured protection for thousands of Hungarian Jews.

Wallenberg worked his kindness at great risk to his own life and has since been memorialized for his boundless compassion and courage. He has been honored posthumously in Israel and in Jewish circles around the world and is remembered with distinction as one of "the Righteous among the Nations."

In addition to Wallenberg, other principled and humane people stepped forward at that time to save Jewish lives. Oskar Schindler, the German industrialist who rescued nearly 1,200 mostly Polish Jews, is probably the best known owing to Steven Spielberg's award-winning film *Schindler's List*. Far less known is Angel Sanz-Briz, a Spanish diplomat, who together with Italian businessman Giorgio Perlasca, procured enough fake Spanish documents to save 5,200 of Budapest's Jews between April 1944 and January 1945, when Russian forces arrived.

As luck would have it, my grandmother, mother, sister, and I were able to obtain Swedish passports. Instead of moving into the ghetto, we moved to a large Swedish-protected apartment building near the Danube. My grandmother was dying. She was already in the advanced stages of leukemia by then but Hungary's hospitals were no longer treating Jewish patients, so of course she moved in with us. Within a week, she passed away.

As a nine year old, I didn't understand much about death. I couldn't have. In retrospect, I also couldn't have understood the gravity of the predicament in which we found ourselves. Knowing that my grandmother's corpse was lying in the next room, I was overcome by an intangible yet profound emotion that left a lifelong impression on me. This much I knew: something sinister had killed my grandmother and left us helpless to remove her body and the evil smell of death it brought. Finally, after two weeks or so, my mother succeeded in bribing a couple of men to come take away my grandmother's decaying body.

As the fall of 1944 slipped into winter, there was little for us to eat and not much to wear. As I remember, every few days soldiers of the Nazi-aligned Hungarian Arrow Cross Party would force Jews to line up on the streets of Budapest and march along the Danube to their executions. Some ten to fifteen thousand Jews were murdered during that period. Somehow, miraculously, Raoul Wallenberg or one of his associates, repeatedly renegotiated for our lives and time after time, together with the other lucky ones, we were allowed to go back home.

In early January, Russian soldiers, who had surrounded Budapest, fought their way into the city. They were our "liberators" in a perverse sense of the word. The atrocities they perpetrated were in many ways comparable to the evil we experienced at the hands of Germans and Hungarians alike. The Russians combed the city plucking from Budapest's streets every able-bodied man they could find, throwing them on trucks to be hauled off to the Soviet Union. Later in life we would meet many Hungarians who would tell similar tales of being captured and forced to toil for three years in Russian labor camps.

In the middle of January 1945, we moved back into the home that had once been the center of my happy childhood. By then my grandparents were dead. Likewise, all of our extended family members who had not been living in Budapest were also dead. (And my father, I would later learn, was wasting away in Mauthausen.) My Aunt Agi, my mother's sister, somehow survived and returned to live with us. Some things were once again the same—just as they used to be. Yet at the same time, some things would never be the same again. It was the beginning of the next chapter of my life.

Decades later I look back with angst and revisit the deep impression left upon me by the curse of helplessness. I never again wanted to be stripped of the power to control my life and shape my destiny.

CHAPTER 2

Growing Up In Fear

Life was chaotic, uncertain, sometimes scary. I found solace in the books that lined the shelves of our heavy bookcases and in the adventures on which those classic tales took me. As the chaos and political turmoil persisted and I grew up, I found a "new normal" in the daily rhythms of gymnasium and university life.

Y OU WOULD THINK that after living the through the uncertainty and chaos of the preceding months, returning to school and the familiar daily rhythms of being a schoolboy would have been a calming, reassuring development in my young life. It didn't work that way. Sure, I was glad to be back in our home with its comforting sights and sounds and smells, but I wasn't interested at all in doing schoolwork.

I did like to read, however. In fact, I *loved* reading. I still do.

I made my way through the wonderful collection of books that lined the beautiful bookshelves in our apartment and my grandparents' apartment, where my Aunt Agi was now living, and was taken in by the works of great 19th and 20th century authors. I was captivated by high-adventure novels, like *The Three Musketeers* and *The Count of Monte Cristo* by Alexandre Dumas, and loved other French greats like Balzac and Anatole France. I read the works of Mór Jókai and Kosztolányi and other famous Hungarian novelists and enjoyed Stefan Zweig, Thomas Mann, and classic German authors.

No doubt about it, I was into books. They took me to fascinating places and interesting times and led me on exciting adventures. As a youngster, I traveled far simply by turning pages and never needed to pack a bag or leave home.

All that reading, however, did little to get me through school. My mother tried to help but it was little use. In any case, she was completely preoccupied at that time just trying to make ends meet. Somehow she managed to reopen our shoe store, but it was difficult to get merchandise in those days. And with a store full of empty shelves, our family business, much like my performance at school, faltered.

I must have driven my mother to wits' end because in desperation she took me to see a child psychologist. I guess he was supposed to figure out what my problem was and whether I had a chance to ever amount to anything in life.

After a few hours of testing, the psychologist reached his conclusion: I had no aptitude whatsoever for science and mathematics, but I had strengths in what today's educators call language arts.

Considering that this fellow knew how much of a book worm I was, and given that I was above my grade level in reading and vocabulary, I would say the psychologist's conclusion was rather predictable. I really hope my poor mother didn't have to pay this character very much for his astute assessment! Besides, now that I think about it, his characterization placed me in the company of more than a few famous scientists—Albert Einstein and British physicist Lord Rayleigh, for example, who as youngsters, were judged to be pretty weak students.

My mother was happy that at least I had some redeeming academic qualities and all hope wasn't lost. Actually, not long after that episode I began to buckle down and started doing the things I was supposed to be doing in school. It was around that time, May 1945, that my father came back home. The miserable and inhumane conditions foist upon him in Mauthausen left his body emaciated, weak, and ravaged by typhoid. But he was back home. It took quite a while, but eventually he regained his strength and health. In time, my parents' retail business began improving—and my school marks did too.

My improved school performance sets the stage for me to tell you a few things about life during my years in high school (*gymnasium* as it was known in Europe) and university. But I didn't just pick myself up and simply waltz off to school every day. Nothing was so simple in those days. In fact, after the Germans left and the Russians arrived, Hungary was in a state of flux. Everything was changing—and in general, not for the good.

The Germans had just ridden out of the country on the famous *Gold Train* laden with carloads of jewelry, gold, artwork, countless heirlooms, and other valuables pilfered from Hungary's Jews. The Germans' departure brought little respite as their Russian replacements tormented Hungary's population—abusing women and forcing men into labor camps. To get some sense of what life there was like in those days, and to construct an image of the chain of events that eventually led me to leave Hungary, let me recount just a handful of highlights to show how history unfolded in the years following the Second World War.

The Communist political movement in Hungary, which was embodied by the Hungarian Communist Party, was alive and well and growing in popularity by 1945. National elections held that year, which were generally regarded as free from voter fraud and shenanigans, were destined to consolidate the Communists' power base—or so they expected. But the election turned out differently. Instead of a Communist sweep, The Independent Smallholders Party captured 57% of the vote, while the Communist and the Social Democrat parties each landed 17%.

The Smallholders' strong showing in the polls was clear evidence that an overwhelming majority of Hungarians wanted parliamentary-style democracy. Hand-in-hand with a vote for that style of government was a desire for landowner rights to private property (which would soon be abolished) and a national economy driven by open markets. The election results were widely viewed as a repudiation of socialism and a state-managed economy.

But with the looming presence of the Red Army and with the assistance of the KGB, the Hungarian Communist Party nonetheless whittled away the opposition through clever political maneuvering and smear tactics. Setting their sights on the Smallholders Party leadership, the Communists went after Zoltán Tildy, Hungary's prime minister in 1945 and later its president, by promulgating allegations of corruption and adultery against Tildy's son. They also forced out the conservative Ferenc Nagy, Hungary's prime minister in 1946 to 1947, indicting him for alleged crimes against the state. As it turns out, Nagy managed to escape to the United States by way of Austria before his trial.

At the same time, the Communist party secretary, Mátyás Rákosi, was propped up as Minister of the Interior, which is critical to this story, because it gave the Communists control of the army and state police. Now the Hungarian Communist Party was on track for purging the opposition from power, absorbing the Social Democrat Party, and shortly thereafter, putting up an uncontested parliamentary ballot.

Here's the executive summary: although the Communists had been soundly defeated in the polls in 1945, in almost no time they had become the country's *de facto* rulers.

So, did all of this history and politics mean anything to my family and me? Absolutely. Under Communism, property and business ownership changed almost overnight. Large privately owned farms were forcibly taken over and replaced with state-managed agricultural collectives. Soon, large companies followed suit, and in time, smaller companies were nationalized. Eventually, even our shoe store was confiscated, thereby wiping out our family income. Before long, our house would also be taken out of our hands.

My parents contemplated leaving Hungary during those times. Around 1948, passports were still available and there were opportunities to emigrate fairly easily. But we didn't. I can still remember the family council—my mother and father, Aunt Agi and her husband Paul, anxiety ridden and solemn—discussing the pros and cons of emigrating. After assessing the details of the situation and its likely resolution, the decision was made to stay put.

The line of reasoning that trumped all others that night—the key argument in favor of staying in Hungary—hasn't left my mind for the past 50 years. Emigrating just wasn't pragmatic, the argument went, because quite simply, our furniture was too cumbersome-, too difficult-, too heavy to move.

Maybe they underestimated the true gravity of our predicament. Or perhaps my family was being overly optimistic—clinging to the hope that things would soon get better. And who could blame them? …considering all that had transpired during that period. It's easy for me to judge now with the considerable benefit of hindsight, but clearly, decisions of life-changing importance sometimes are made for the wrong reasons.

Being too attached to material possessions—like beautiful dining room furniture and the carved bookshelves that held the books I loved so much, can mean the difference between life and death. It's hard to know for sure when these ideas formed in my mind, but I think as a result of that experience, I decided never to allow myself to be held captive to material comforts—never to let worldly possessions dictate my thinking when it comes to life's important decisions.

There's plenty more I could tell you about the changes that overtook Hungary when I was a youngster, but I think I have told you enough to paint a picture of the conditions and mindset that prevailed at those times. It was against that backdrop that I made my way through the Hungarian high school (gymnasium) and university educational systems.

Even amidst the political upheaval that racked Hungary, its institutions of learning remained world class. I had the good fortune to attend the famous Minta Gymnasium on Trefort Street. That was a school that produced some of Hungary's best and brightest, including Edward Teller and Theodore von Kármán, the aerospace engineer who helped establish California's Jet Propulsion Laboratory. Of course, I didn't know Teller and von Kármán from school—they were a generation older than me. Just to mention, by wonderful coincidence, von Kármán was honored in 1963 with the first National Medal of Science. I received that honor in 2002.

Due to the changing political environment, Hungarian high schools were gradually instituting a Socialist education curriculum. That meant that students in my year learned Latin and Russian simultaneously. It also meant that I had to read and memorize the history of the Bolshevik Party, the forerunner of the Communist Party in the Soviet Union.

Now, because of my background—an upper middle class Jew whose father had been a bank director (until he was sacked), I was acutely aware of my *bourgeois* status. Call it what you like—class enemy, undesirable, etc, I was *persona non grata* and always felt like I was skating on pretty thin ice. So I needed to excel in these Socialist subjects to avoid getting in trouble at school—and I did.

Don't get me wrong. It wasn't all gloom and doom in those years. Human nature has a way of helping people find comfort and pleasure in tough times and, I certainly have some fond memories from my teenage years. In the classroom for example, I had an excellent Hungarian language and literature teacher, who made the material interesting and enjoyable. The same was true for mathematics and physics. And although just a short while earlier, a child psychologist had pronounced me wholly incapable of learning those subjects, I actually enjoyed them and did rather well.

Outside of the classroom, I developed a love for sports—especially basketball. Now, I know there are a lot of people today on the U.C. Berkeley campus and quite a few other college campuses for that matter, who would be pretty surprised to learn that once upon a time, I had a pretty decent jump shot. But it's true. I wasn't half bad at defense either! Before too long I became a starter on our high school team (usually playing point guard, the guy who sets up the plays) and was heavily invested in our team's success. We played against other schools in Budapest and it was all very exciting, especially when we played our main rivals in the big games that drew large crowds. My father was always there cheering me on—I felt so proud when I would see him in the stands—and my girlfriends came too.

I was an enthusiastic basketball player in my high school and university years. I stand second from the right, number 8, in this photo from 1952.

I had a fairly typical—if active—social life as a teenager, and I did the sorts of things teenagers do (or at least did back then). Much of the socializing was tied to school and sporting events, but not all of it. We used to play this "finger soccer" game by shooting polished buttons with our fingernails across a tabletop. My male friends and I took this little game of ours kind of seriously and even held playoffs and tournaments.

And like a typical teenage boy, I socialized with girls when I had the opportunity to do so—after the basketball games, for example, and outside of school. I used to try my utmost to earn a little pocket change so that I could take a girl out for a cup of coffee or go dancing in the evenings or at a weekend party. I can still remember a strategy my friends and I concocted so that we could do a little drinking at these parties without becoming too tipsy. I used to drink a half cup of melted butter to coat my stomach in order to slow down the diffusion of alcohol through the stomach walls into the blood stream. I can tell you from first-hand experience that the procedure actually works. But as I look back on it now, I think the idea of drinking half a cup of melted butter for any reason is pretty disgusting!

That was about the extent of the fun. By this point in my high school career, the Communists were in full political control of the country. Agriculture, finance, manufacturing, and all import, export, and natural resources were controlled by the Communists. Hungary was being pushed to invest intensely in heavy industry largely because doing so benefited the Soviet Union. As a result, production of metal products and machine parts rose quickly, while food and consumer goods—clothes, shoes, even soap, were all in short supply.

What was plentiful in those days was paranoia. People were afraid to talk to one another—to complain of the difficult circumstances and the way Hungary was being ruined—for fear that someone listening, some neighbor or "friend" would pass the word along to the AVO, the hated secret police, KGB-equivalent in Hungary. Nowadays, the torture and terror tactics employed by the AVO are open to the public for viewing. The AVO headquarters on Andrássy Avenue is a museum known as The House of Terror.

But back then, when word reached the AVO that someone made a practice of criticizing the State, the consequences were dire. People we knew disappeared—snatched from the streets or their homes, perhaps never to return, or maybe to return years later broken in spirit and body, never speaking of what transpired while they were gone. We know now that so-called enemies of the State were rounded up and sent off to secret internment camps dispersed throughout the country. Records show that in May and June of 1951, almost 13,000 upper- and upper-middle-class people were taken from their homes in Budapest and deported to internment and labor camps. We were living in fear.

Publicly, anyone who knew what was good for him attended the meetings and pep rallies that glorified the Communist Party and clapped enthusiastically when Party members' named were mentioned. We feared that whoever stopped clapping first would likely be suspected of being an enemy of the regime. So we plastered fake smiles on our faces, let their propaganda wash over us, and clapped and clapped. But privately, we drew our curtains closed every evening at 8 o'clock and tuned in to BBC and The Voice of America to learn what was really going on—in Hungary and the world at large. We all knew that everyone did it. But for fear of repercussions, no one said a word.

And so it was that in 1953, in the darkest period of Stalinism, shortly before Joseph Stalin died, I graduated from high school. At that age, I truly loved to read and write and was endlessly fascinated by history. And so I decided that I wanted to become a historian. My father was alarmed with my career choice, and for good reason. I would later learn by reading the biography of Eugene Wigner that the conversation that unfolded between me and my father that day, had taken place almost verbatim some 35 years earlier between Wigner and his father. When Wigner told his father he wanted to be a physicist, the senior Wigner asked "How many jobs for such physicists exist in our country?" The answer was "four."

Similarly, my father sat me down and explained that a small country like Hungary does not need too many historians—and my prospects for making a living in that profession were pretty poor. However, if I became a chemical engineer, my father advised, I could make a good living. Chemical engineers were needed the world over, he reasoned.

It was a profession that could be practiced most anywhere. As it turns out, Wigner's father had made the same suggestion to his son. I took my father's advice and agreed I would study chemical engineering but decided I would always continue to read history.

The question now was how to get accepted to a university. I finished high school at the top of my class and received the Rakosi Medal, an honor awarded to each year's valedictorian. But academic merit meant little given my *persona non grata* status. No one wanted *bourgeois* students on campus. So my father, being the pragmatic person he was, hatched a plan to get me accepted on account of my basketball skills. I don't exactly know how it happened, but through connections and bribes and favors and payouts, my father managed to persuade my basketball coach to persuade his friends to make the case for accepting me as a chemical engineering student at the Technical University in Budapest.

My early education in chemical engineering and chemistry took place at the Technical University of Budapest, where I studied from 1953-1956.

In those days, about 200 students were admitted to the program but only 50 would graduate because according to the terms of our state-planned economy, no more than 50 chemical engineers would be needed four years down the road. Thus the pressure to succeed in school was enormous.

As a new student in chemical engineering, my studies were difficult but I enjoyed them. For example, I remember mathematics being challenging but in an intellectually rewarding sort of way. Likewise inorganic chemistry was also enjoyable, especially because the professor liked poking fun at the teaching assistants who ran the demonstrations and experiments. I remember one course in particular, in crystallography, because it called for a ton of brute-force memorization. For the exams, we had to memorize all the crystal systems—orthorhombic, tetragonal, hexagonal—and be able to quickly spit back all their symmetries and structures. I passed the course with flying colors but then promptly forgot half the material I had memorized.

About a year after starting at the university, I was at a party where I met Judy Kaldor. From that point on, she was my only girlfriend. Eventually we left Hungary together and got married—I'll tell you more about that later—and after more than five decades we are still husband and wife.

Judy's friends soon became my friends and she began showing up regularly at my basketball games. The opportunities for socializing multiplied quickly but I found that with all the school work I needed to do, I was just too pressed for time to continue playing on the team. So I gave it up. Not only did I need to excel in the technical courses, I had to perform well in all the coursework in Marxism-Leninism that remained mandatory throughout the time I was in school.

One thing that stands out in my mind today about my university education is that examinations were always oral. At the end of the semester, we had to stand in front of the professor and answer any kind of unexpected question that was thrown at us. Some of my friends found the experience positively nerve wracking. But I actually liked these examinations.

I quickly learned that by watching the professor's body language I could more or less guess if I was on track to the correct answer. When I saw I wasn't heading in the right correction, I changed course—always keeping my eye on the professor for a clue as to how well I was doing. I was doing just fine. Three years into university my grades were high and I was sure I would wind up with the select group of 50 students who would earn a diploma and be eligible for a job.

After becoming reasonably confident that I knew how to manage my schoolwork, I began working on strategies for raising a little pocket money so that I could afford to take my date to dinner or for an evening at the opera or theatre. Here is where a little chemistry know-how came in very handy.

I used to concoct liqueurs by fermenting sugar solutions and adding various flavorings—and then sold the goods to friends of the family. I also ran a brisk trade in dyed stockings. As I mentioned, consumer goods were in short supply at that time and women's stockings certainly weren't easy to come by. People used to collect nylon thread from worn and frayed stockings and reweave them. But the recycled stockings didn't look as good as new ones. So I experimented making batches of colorful dye solutions and learned how to spruce up the recycled products.

Another one of my money raising schemes leaves me scratching my head when I think about it today. I used to sell mercury thiocyanate "snake eggs" for children to play with. The material is white but when lit by a match, it evolves a thick black smoke that looks like a snake rearing up from the snake egg. Such a toy just wouldn't fly today because of its chemical toxicity but I didn't think much about the hazards at the time and the product sold pretty well.

Another "extracurricular" activity during my university years was military service. University students typically attended six weeks of military camp during the summer in order to avoid being drafted into the army. None of us liked it. Army officers looked down on students. We were weak and wimpy as far as they were concerned and they did their best to prove it by testing our endurance and strength with impossible demands that ranged from physical hardship to emotional trauma. Overall the experience was downright unpleasant. The only saving grace was that it was over in six weeks.

It was so maddening to see how the army operated under Communist rule. It was clear that the Hungarian military was cold to the Communist leadership and they were doing what they could to push back and call their own shots. Amongst our fellow students—when we were certain no one was snooping—we used to say that if Western powers wanted to penetrate the Iron Curtain, Hungary would be the ideal place to do so. We were certain that our army would never lift a finger to oppose a Western force that sought to push out Communist rule.

The reign of fear that had terrorized Hungary for years slowly began to subside following Stalin's death in 1953. Stalin's successor, Nikita Khrushchev, espoused a different view of Communism and sought to do away with some of the former leader's policies. The news from Moscow affected Budapest and Hungarian politics soon began changing. People disappeared less frequently, cracks began appearing in the terror network, and soon citizens began talking more openly. For the first time in years, Hungarians began criticizing the government in the form of jokes and songs and humorous performances. And it was all tolerated.

Hungarians hated the repressive regime and began sowing seeds of discontent that emboldened us to speak out ever more openly and begin demanding freedom. Men, women, workers, students, and the country's *intelligencia*—young and old alike—were thirsting for a better life and growing strong enough to say so. These were the conditions that precipitated one of the most transformative events of 20th century Eastern Europe: the Hungarian Revolution.

CHAPTER 3

Revolution and Escape

*As quickly as it began, our Hungarian Revolution was crushed.
There was no time for contemplation, second thoughts, or
contingency plans. Like thousands of our countrymen, we
fled on impulse. We trekked in darkness and fear—and
through great fortune, found our way to freedom.*

NEVER IN MY wildest dreams would I have guessed that life
was about to change drastically and irreversibly. All things
considered, in the fall of 1956, life was pretty good and getting
better. I was doing well in school and just starting the fourth year of
my chemical engineering education. I led a happy and active social
life that was centered about Judy and our growing circle of friends.
And the improvements on the political front—namely the general
tolerance by the Communist regime of open criticism—certainly

engendered a sense that things weren't too bad and that they would only continue to improve.

They did—at least for a while.

The anti-Communist jokes and slurs that Hungarians exchanged at this time gradually gave way to disparaging songs, which in turn led to the humorous critical plays and performances I mentioned previously. All of these actions and the apparent absence of any official counteraction bolstered our spirits and encouraged us to take the next step—to demonstrate openly and collectively.

And so it was that groups of Hungarians—workers, students, and others—began raising their voices and demanded to be heard by the Soviet-controlled government. Some of the older students, myself among them, posted signs around the campus of the Technical University announcing that on October 23rd we were going to march and hold a public rally in support of our freedom from Soviet occupation. We were also gathering to publicly declare our solidarity and sympathy for Polish citizens, whose political struggles at that time and earlier in history mirrored the ones endured by Hungarians.

On that unseasonably warm autumn day, our group assembled at the gates of the university—lining up by department and class—and marched through the streets from the university campus to the Hungarian National Museum. There, at the very place where in 1848 the nationalistic Hungarian poet Sándor Petőfi delivered his rousing freedom speech, we began chanting the lines of his famous poem, "Talpra magyar, hív a haza!," which loosely translates as "On your feet, Magyar (Hungarian), the homeland calls!" We also aired our grievances and demanded an end to Soviet occupation, free and fair elections, and other political freedoms.

Meanwhile groups of workers and others rallied in another demonstration, this one in front of a huge statue of Stalin, which they promptly and boldly began dismantling. Eventually the statue was toppled and the crowd celebrated with the now famous symbolic gesture in which they draped Hungarian flags in the only part of the statue left intact—Stalin's boots. At the same time, other groups demonstrated near the Parliament building and in front of the Hungarian radio station.

At first all of the demonstrations were peaceful. But soon, as night fell, the hated secret police, the AVO, appeared on the scene at the radio station and tried to quash the demonstration. Soon fights broke out and the AVO fired upon the crowd, killing several demonstrators, and touching off a riot that set the Revolution in motion. That evening, the Hungarian army units stationed in Budapest joined the revolution—bravely siding with the demonstrators. By then a mass of protesters, largely indistinguishable as students, workers, or other Hungarian citizens, had taken to violence. The group destroyed symbols of the Communist regime and took control of military depots, widely distributing guns to their able-bodied countrymen.

In the fall of 1956, Hungarians, like the ones shown here near a captured Russian tank, celebrated their newly-won freedom, which turned out to be very short-lived.

Under cover of night, Soviet tanks entered Budapest, but the armored units were reluctant to fire upon the citizenry. The city was transformed into a war zone. The police disbanded and the AVO tried to disappear. Yet several AVO members, trying to hide among the crowds, were recognized, apprehended, and hanged for the terrible suffering they had inflicted on their fellow Hungarians.

The violence continued into the next day and sporadically for a few more days thereafter. Small pockets of Hungarian resistance fighters took on Soviet forces, battling them with confiscated munitions, Molotov cocktails, and anything else they could get their hands on. Gradually the fighting quieted down. By October 28th a general cease fire went into effect and by the 30th, much to our joy, Soviet troops had largely retreated from Budapest to the countryside.

As those events played out, university students quickly recognized the need for organizing makeshift police units to maintain law and order. We dispatched units to guard key university buildings and other important sites throughout Budapest. We all had machine guns—they had been secured days earlier with the help of the Hungarian Army. But these weapons were mostly for show—to keep some semblance of civility amongst the population, to prevent looting, and avoid outbreaks of violence.

In a curious twist of fate—just one among many that would shape my future in the days to come, my friends and I drew up lists of student participants—young adults who were involved with various aspects of the Revolution and who could be called on in the days ahead to assume responsibilities and manage important tasks. The drive to organize and distribute printed lists came to us rather naturally as Hungarians and university students. The trouble was— these lists, which included my name and those of my friends, would all too easily fall into the hands of government officials, making it fairly simple for them to track us down.

But for the time being, everything looked bright. Soviet troops were withdrawing, and as far as we knew, they were leaving for good. At the same time, Imre Nagy, the Hungarian prime minister, abolished the AVO and was in the process of forming a just and representative government. Hungary was clearly poised to withdraw from the Soviet bloc.

For the first time in years, the borders to the west were opened widely. Food, clothing, and other goods that had been in short supply came pouring into Hungary via Austria. International aid came too, and with it, reporters and journalists to document our triumph. We were ecstatic! These were happy times for Hungarians. The nation's spirit was unimaginably high and the road to our future was shining brightly.

But the lights went out suddenly. Hungary's short lived freedom ended dramatically early on the morning of November 4th, 1956. Without warning, all hell broke loose as an overwhelming Russian force with rows of tanks and armored cars rolled into Budapest to put down the uprising once and for all. As it happens, I had left my machine gun in my locker in the organic chemistry laboratory and couldn't get it. It wouldn't have done me much good anyways. What could I have done to drive back a Soviet armored division? As it was, the little bands of resistance that cropped up here and there were quickly squashed. Still, the thought has remained with me all these years that when the Russians came back to take Budapest, I didn't have a gun.

As the situation deteriorated and Hungary quickly fell into Soviet control, the anti-Communist support and encouragement we had been receiving in recent weeks from the U.S. and other western countries—much of it by radio broadcast—began to vanish. Unfortunately for Hungary, much of the West's attention had suddenly been diverted from Hungary to the Sinai Peninsula where Egypt had just nationalized the Suez Canal, touching off a major Middle East crisis.

Hungarians were left to fend for themselves and it was just a short time before the streets were abuzz with news of widespread arrests. Nationally known figures including Prime Minister Nagy and Pál Maléter, the colonel of the Hungarian armored division who valiantly opposed the Soviet forces, were among the first to be taken. They were later executed. But by November 20th, the news turned far more personal as we learned that some of the student activists from the Technical University had also been arrested. Ultimately I was to learn that of the 50 students left in my class after four years of being weeded out academically, more than half either died during our failed revolution or escaped the country.

All the while, anxious conversations crisscrossed Budapest's telephone lines. Word spread quickly that the police had begun raiding houses within our circle of friends. They were searching for those of us who were unlucky enough to have been named on those fateful lists that we had drawn up just days earlier—back when we were full of hope.

In quick succession, my mother received two alarming phone calls from the mothers of my friends warning us that the police had just been to their houses looking for their sons. It was clear that I needed to leave—and do so immediately.

That night my parents and Judy's parents decided that Judy, my sister Marietta, and I would leave Hungary the next morning. In hindsight, the decision was courageous all around. Just imagine, well established families encouraging their young adult children (Judy was just 18) to escape from their home country—alone—during uncertain and dangerous times. They were bold decisions indeed, yet they were made without a moment's hesitation. Everyone knew it was the right thing to do. Judy has long said that the greatest gift her parents ever gave her was permission to leave Hungary with me.

Most of the transportation services, like other services and businesses around the city, weren't running at that time. But some westbound trains continued to operate and the border to the west remained open perhaps because the Russians were willing to allow the least content Hungarians—potential troublemakers—to sneak out of the country.

The next morning, roughly the 20th of November, both families met at the train station in Budapest. With little fanfare but much angst, the three of us were sent off with our parents' blessings and well wishes. We climbed aboard the westbound train not knowing our fate but acutely aware that we sat among carloads of escapees hoping for something better in the west.

Not two hours into the journey the conductor came to warn us that a Russian patrol was waiting at the next station ready to arrest anyone they saw fit to arrest. The train stopped in the middle of nowhere. "Anyone who wants to leave," the conductor advised us, "now's your last chance to get off the train freely." Like most of the refugees on that train, we had no travel permits, no papers documenting our trip. With little choice, hundreds of people disembarked and began walking westward following the direction of the railroad tracks.

There we were in rich farmland. Soggy, muddy, November farmland—trudging through the Hungarian backcountry at dusk in shoes that grew heavy with muck as we walked on in the waning light. It wasn't long before darkness settled upon us and my sister insisted she couldn't go any further. So I fell behind with her, helping and pushing her forward and at the same time, losing Judy in the darkness. Marietta and I soon spotted a little village, and as luck would have it, just as we reached this little hamlet, another group from the train—among them Judy, also arrived there.

We desperately needed help to make our way to the border and so we took a chance by knocking on the door of one of the farmhouses where we saw light. Luckily, the family was sympathetic. We scraped together whatever money we had and offered it to the farmer's son to be our guide—to see us safely to the border. He accepted.

Now, with the benefit of this local's knowledge as to where the Russian searchlights were positioned and which of them were working and which were not, we walked on in the darkness. Moving from haystack to haystack for cover from the searchlights and a bit of shelter from the cold, we hiked through the fields at night, hiding during the daytime, slowly pushing on toward the border. Along the way, we met another group—this one with a few small children and a baby—and continued westward with them, slinking through the darkness, nipping occasionally at a bottle of rum that Judy's father gave us to bribe soldiers but which instead gave us a feeling of warmth.

Around 2 AM on the second night of walking, the border to Austria came within sight. The farmer's son had done his part and now he turned to go. Freedom and security lie just a short silent walk ahead but so did search lights and Soviet troops. What's more, due to the zigzagging shape of the borderline, it wasn't at all obvious which path would take us to safety and which would undoubtedly lead us to being arrested. And then as though the situation weren't tense enough, our group's silence was suddenly shattered as the baby began crying! It was sheer panic. The Russian troops had surely heard us—or did they? By an incredible stroke of luck, it turned out that they had not heard us. Breathing a collective sigh of relief, we steeled ourselves against a thousand fears and finally figured out how to navigate the final stretch to the Austrian border.

In the end, we made our way to a Red Cross station on November 27th or 28th, 1956. The station was manned by Austrians and others who had come to help Hungarian refugees. After a brief rest and hot drink, the three of us boarded a bus for Vienna. I paid for the bus tickets using money my father had stashed away for us. As I recall, before we left Budapest he had emptied a tube of toothpaste, opened the wide end of the tube, and inserted a U.S. $20 bill inside for safekeeping. I extracted this cleverly hidden bill, exchanged it for Austrian schillings, and bought our bus tickets.

Soon we were in Vienna, where we found stations that had been set up to help refugees with their basic needs. Judy and Marietta went to a local cloister where the nuns looked after them and other women. I made my way to a station run by the World University Service of the International Rescue Committee and from there to a student fraternity or "studentenheim" where male refugees were given food, lodging, and shelter. I can't recall now what exactly was going through my head as I finally laid down to rest. But being the forward-looking person that I am, I was probably thinking about how we would make our way to the United States, which was where I decided I wanted to go.

With that plan firmly in mind, I headed over to the American embassy the next day to inquire about immigration. Judging by the throngs of people standing in an endless line, it seemed that everyone wanted to go the U.S. and our chances of making it there seemed bleak. But with another stroke of good fortune, I learned that the World University Service was able to arrange for university students to be granted priority status giving them a major advantage with regard to U.S. immigration quotas. As luck would have it, both Judy and I qualified for this preferential treatment and our names were soon entered on this top-priority list of U.S.-bound refugees. It was also around that time that Judy and I became engaged. By the way, more than 50 years later, we still wear the very same wedding rings we bought at that time in Vienna.

As for Marietta, let me jump ahead and tell you that on my parents' suggestion, she made contact with my mother's brother who lived in Santiago, Chile, and ended up moving there. My parents would later follow her to Chile. Eventually they would all end up settling in the U.S., but I'll tell you about all that later.

Meanwhile, back in Vienna, around December 1st, I spotted a notice on a bulletin board in the studentenheim announcing that Cornelius Tobias, a Hungarian who was a professor of medical physics at the University of California, Berkeley, was coming that evening to talk with students interested in going to the United States. Tobias had been visiting at the Karolinska Institute in Stockholm when the Hungarian revolution broke out and he had come to Vienna to see if there was something he could do to help.

I arranged to meet with Professor Tobias and we spoke privately. When I told him that I was a chemical engineering student at the Technical University, he quickly offered that his brother, Charles Tobias, was a professor of chemical engineering—also at U.C. Berkeley. He gave me his brother's address and suggested that I contact him, saying that perhaps with his connections in chemical engineering, Charles would be able to help me continue my studies at Berkeley once I made my way to the United States. I didn't know if this was a long shot or not, but I certainly wasn't going to delay writing the letter.

A few days later, the students on the first-preference list were moved to another residence where we stayed until January 2nd. After a New Year's party for refugee students at the Rathskeller in Vienna, we traveled by bus to Munich and then on to a U.S Army base a few hours away. We stayed at the base for two days and then boarded an Air Force plane for our first ever plane ride, which took us to Newark, New Jersey. From there we ended up at Camp Kilmer, a World War II Army base that had been converted to a refugee camp to house émigrés fleeing from Hungary.

Just one day after arriving at the camp, Judy and I received our green immigration cards. I immediately started looking for a job. Within a week, I moved into the International House at Columbia University and Judy moved in to the dormitories of New York University in Manhattan. That same week I also received a reply from Professor Tobias. Enclosed with his letter was a letter from Professor Kenneth Pitzer, who at that time was Dean of Berkeley's College of Chemistry, which to this day includes the chemistry and chemical engineering departments. Unbelievably, Professor Pitzer and Berkeley were offering me a teaching assistantship "on probation" as an entering graduate student and granting admission to Judy as a freshman in the College of Chemistry.

Charles W. Tobias (left) and Kenneth S. Pitzer, Dean of the College of Chemistry, at the University of California, Berkeley.

Then on January 22nd, 1957, we received two one-way airline tickets courtesy of the World University Service and boarded a plane to San Francisco. There at the San Francisco airport, Professor Tobias kindly greeted us and took us to Berkeley. He had arranged for me to stay with a lady in Berkeley Hills who rented rooms to students and also arranged lodging for Judy with another family on the same street. He was also kind enough to secure loans from two other couples to pay our tuition fees for the spring semester so that we could be admitted promptly and begin our new lives as U.C. Berkeley students without delay.

More than 50 years after those life changing events, I still recognize that at every twist and turn during the long journey from Budapest, anything could have abruptly caused our story to come to an unhappy ending. During the most chaotic moments, we didn't have time to mull over our options. We had to quickly decide on a course of action and follow it. And we were fortunate that things turned out well. We know that we were among the lucky beneficiaries of a number of kind people and of a new country that we would soon find to be welcoming and gracious. We remain forever grateful.

UNIVERSITY OF CALIFORNIA

DEPARTMENT OF CHEMISTRY
AND CHEMICAL ENGINEERING
BERKELEY 4, CALIFORNIA

January 8, 1957

Mr. Gabor Somorjai
WUS [IRC]
Camp Kilmer
New Brunswick, New Jersey

Dear Judit and Gabor:

First of all, let me greet you warmly on the occasion of your arrival to this blessed country. I hope that you will be as little disappointed in your expectations as I am after 10 years of life here.

Now let's get down to business. (1) The Dean of the Graduate Division has agreed to admit Gabor as a graduate student in chemistry or chemical engineering (see enclosure I) at the University.

(2) The Dean of the College of Chemistry accepted Gabor as a graduate student in chemistry and offered a very generous stipend of around $800 for the next 5 months beginning February 1 (see enclosure II).

(3) The Admissions Office of the University agreed to admit Judit to the College of Chemistry as a special student (see enclosure III). As to the exact level at which Judit will continue her studies; this will be decided after you all arrive here and will have a chance to talk to the professor who is in charge of advising students. It is very probable that neither of you two will suffer any loss of time due to the revolution and the months of darkness following it.

You need not be concerned at this time about exactly what degree (M.S. or Ph.D.) will Gabor want to take. You will have plenty of time to familiarize yourself with the local system of education and the future possibilities after you have arrived in Berkeley.

The University quite probably will waive the non-resident tuition fee in the case of Gabor because of his outstanding record and his status as a graduate student. However, we will have to pay a non-resident tuition fee ($150) for Judit. If the WUS or the IRC is willing to provide this, the better. If not, Gabor will have to pay it out of his stipend.

Mr. Gabor Somorjai   - 2 -   January 8, 1957

Now as to living quarters, food and incidentals. (1) A Hungarian student at this University who is married to an American girl and who lives close by the University offered to take in Judit for a period of six months. (2) Gabor can well afford to rent a room and pay for his food from the stipend offered by the Dean. However, I have also found another family who would be glad to furnish him with a room and bath plus food for some work to be done around the house. Whether Gabor will want to rent or stay with the family can be decided after you arrive in Berkeley.

I trust that the letters (I, II, III) and this letter will be adequate insurance to the authorities and they will make immediate arrangements for your transportation. We will receive you at the airport whenever you arrive and put you up with friends as long as you have not settled finally. However, if the authorities require further guarantees (sponsorship?), please write airmail immediately so that we can find two sponsors and facilitate your speedy arrival. The reason why you ought to make every effort to arrive at Berkeley as soon as possible is that the spring semester begins on Monday, February 4, and you should try to be here if possible at least a week before this date so that your program could be worked out, you could register, and settle down. Naturally, even if it were impossible to arrange for your arrival before this time, you needn't fear that your case is lost.

It was my understanding that the IRC will provide the transportation costs and even possibly a sum necessary to start your student life at the college of your choice. (If this is not the case, we will still find means to pay for Judit's registration and to cover your initial expenses). The most important thing at the moment is to get you out here without delay.

Forgive me for not going into further detail; we will have plenty of time to discuss other problems after you have arrived in Berkeley. With fondest greetings, I am

Sincerely yours,

Charles W. Tobias
Associate Professor of
Chemical Engineering

Enclosures

Letter from Charles Tobias

COPY

UNIVERSITY OF CALIFORNIA

DEPARTMENT OF CHEMISTRY
AND CHEMICAL ENGINEERING
BERKELEY 4, CALIFORNIA

January 8, 1957

Mr. Gabor Somorjai
WUS/IRC7
Camp Kilmer
New Brunswick, New Jersey

Dear Mr. Somorjai:

I am glad to inform you that you will be accepted as a graduate student in this department, beginning with the spring semester 1957, to study either chemistry or chemical engineering, as you choose.

I am also glad to be able to offer you a fellowship of approximately $825 for the spring semester.

If your record with us is as outstanding as it has been in the past, you may feel assured that we will be able to continue to award you some sort of a fellowship in order that you may complete your studies.

Our spring semester begins on January 28 and it would be well for you to plan to arrive a few days before that date in order to complete your registration.

Sincerely yours,

K. S. Pitzer, Dean

Letter from Dean Pitzer

CHAPTER 4

Chemistry in the New World

1950s America welcomed a young European couple with open arms. The Land of Opportunity and Berkeley, California, gave us a new home, an education, and jobs. Most importantly, it gave us a sense of belonging, security, and peace of mind.

A T THE TIME, I didn't know what it meant to be "green" and I hadn't yet learned the expression "just off the boat," but I certainly understood what the experience was all about. As brand new immigrants to a country far from our homeland—as newcomers to a distant place with unfamiliar ways, curious customs, and a somewhat incomprehensible language, Judy and I were as green as could be.

Sure, we had both studied English privately in Hungary, but those kinds of lessons can't prepare someone who's suddenly dropped into a new country to understand the rapid-fire barrage of new words, idioms, and expressions raining down on all sides. Internalizing the ocean of *written* text was challenging enough—but *spoken* English was especially hard to comprehend because everyone seemed to be talking so quickly. We certainly weren't prepared to keep up with chemistry lectures!

But there we were in Berkeley's College of Chemistry—Judy, a freshman, and I, a new graduate student with probationary status as a teaching assistant, and we didn't have much choice. That's just how it is for immigrants; sink or swim. We swam.

We enrolled in English classes for foreign students and jumped head first into *Americana*. The first couple of months were especially difficult. In addition to a steep language barrier, we also had to get used to a new style of examinations. I mentioned earlier that exams in Hungary were oral, and if you were a reasonably good student of science and sufficiently astute to monitor and interpret the professor's body language, it was easy to sail through those exams and get top marks. But in the States, the exams were written—and they were quantitative, that is, you had to do a lot of mathematical problem solving—and to get good grades, you had to do it quickly. Somehow we managed. In fact, we both did very well in our first semesters.

I wasn't at Berkeley very long before I needed to make some important choices regarding the direction of my education. I had to choose between chemistry and chemical engineering. Of course there is plenty of overlap between the two fields. Students in each area study general-, organic-, physical chemistry, and other topics, but eventually the course work becomes specialized.

In Hungary and much of Europe at that time, an education in chemical engineering was more focused on scientific aspects of chemistry than on hard-core engineering concepts. Eventually chemical engineering students everywhere have to study quantitative methods for evaluating mass transfer and heat flow and topics such as computational fluid dynamics, for example. But I wasn't especially interested in those subjects or in mathematical methods for designing advanced chemical reactors. Of course they are important; I just wasn't excited about them.

I recognized that I was more interested in the science of chemistry than in chemical engineering but I wasn't sure which of the chemistry disciplines interested me most. That all changed when I took my first graduate-level course in thermodynamics. The course was taught by Leo Brewer, a fantastic lecturer and, as I soon came to learn, a great scientist and wonderful person.

Leo Brewer

Leo Brewer was a world expert in high-temperature chemistry of inorganic compounds. During the Manhattan Project, he played a crucial role leading a research team charged with determining the best means and materials for handling molten plutonium. Through an exhaustive analysis of numerous elements' high-temperature chemical and materials properties, his team came up with the answers and in the process discovered a number of important classes of materials with superior high-temperature characteristics. Among them are the so-called refractory sulfides of barium, cerium, and thorium, and the mixed cerium-thorium and uranium-thorium sulfides. One key result of his work is that Los Alamos National Laboratory made several hundred crucibles from these materials and had them on hand and ready for use when they were needed to contain the first samples of plutonium.

Sadly, Leo suffered from his World War II work with beryllium, now categorized as a carcinogen but which was mistakenly thought to be benign at the time. However, he was certainly full of life and enthusiasm when I took his thermodynamics course and he was very much my inspiration for choosing a path in chemistry and in particular, in physical chemistry. He would also later turn out to be a pivotal figure for me in another way; by helping me get my footing in the research world.

So with my mind made up to follow a path in chemistry, I began to establish myself by the end of that semester, as a diligent and conscientious young scientist. At that point, I was accepted as a full time graduate student in the chemistry department at Berkeley's College of Chemistry.

The University of California, Berkeley, was an exciting place for an enthusiastic young scientist to be in 1957. As one of the main scientific powerhouses during the Manhattan Project era, Berkeley's reach, notoriety, and expertise extended beyond the Radiation Laboratory, now known the Lawrence Berkeley National Laboratory, to the other national labs, including Los Alamos, Oak Ridge, and the University of Chicago's Metallurgical Laboratory.

In the run-up to the Second World War, numerous world-class scientists plied their trades on the U.C. Berkeley campus. Among the most famous were J. Robert Oppenheimer, the theoretical physicist who would later be known as the "father of the atomic bomb," for his leading role in the U.S. nuclear weapons program; Ernest O. Lawrence, who established Berkeley's Radiation Laboratory, later renamed in Lawrence's honor; and the charismatic Edward Teller, a fantastic lecturer who championed the hydrogen bomb and nuclear energy, and who cofounded the Lawrence Livermore National Laboratory.

Also at Berkeley in those days was Glenn T. Seaborg, the Nobel-Prize-winning nuclear chemist who was the principal or co-discoverer of ten (transuranic) chemical elements. One entry on that list is plutonium, the element for which Glenn Seaborg developed the chemical separation methods used to prepare the fuel for the world's first plutonium bomb.

Gilbert N. Lewis (left), Glenn T. Seaborg (center) and George C. Pimentel (right) were instrumental in developing UC Berkeley's reputation for excellence in physical chemistry research.

But the foundation for Berkeley's excellence in the physical sciences, and in particular in physical chemistry, dates back to an even earlier era and to another seminal physical chemist, Gilbert N. Lewis, who served from 1912 to 1941 as the College of Chemistry's first dean.

Lewis' legacy was multifaceted, to say the least. He was an original thinker who contributed substantially to modern chemistry's understanding of covalent bonds and the role of pairs of electrons therein. He helped develop methods for accounting and explaining the thermodynamics of chemical systems—meaning the interplay of chemical reactions with heat, work, and energy. He also developed a novel conceptual framework for understanding acid-base chemistry based on the electronic properties of reactants that every student today knows as "Lewis" acids and bases. Lewis also pushed forward the field of light-driven or photochemistry and our understanding of chemical isotopes, such as deuterium, the heavy part of "heavy water."

But he did more than all that. As part of his negotiations to come to UC Berkeley from Massachusetts Institute of Technology in 1912, G. N. Lewis strongly lobbied to establish the College of Chemistry as an independent academic unit responsible for reporting directly to the university president—not a department within a College of Arts & Sciences or similar university entity. As a result of that organizational structure, which still exists today, the chemical sciences at Berkeley were given a special opportunity to thrive, build an outstanding reputation, and attract scientific talent.

Lewis also had the foresight to procure funds to build a machine shop—a facility almost never found at that time in academic chemistry institutions. Ready access to the machine shop set a precedent nearly 100 years ago for faculty members to design and build their own customized laboratory equipment to carry out experiments. That form of experimental creativity would one day become a cornerstone of my own research program.

Above and beyond all else, Lewis was especially committed to mentoring young scientists. His devotion to pedagogy gave rise to a generation of scientific leaders, such as Seaborg, who, after serving as a post-doctoral chemistry researcher with Lewis, carried on the senior chemist's traditions. Chief among them was encouraging researchers in related but traditionally separate fields to bring together their distinct expertise to tackle major scientific challenges. One extensive example of such a fruitful interdisciplinary collaboration was noted by Seaborg at the ceremony marking his receipt of the 1951 Chemistry Nobel Prize, which he shared with Berkeley physicist Edwin M. McMillan.

Seaborg pointed out in his Nobel address that the discoveries in the field of transuranium elements were "made possible by...the unusual and excellent spirit of cooperation which exists at the Radiation Laboratory," the forerunner of the Berkeley Lab established by Ernest O. Lawrence. In setting up that facility, Lawrence himself made a concerted effort to recruit scientists with disparate backgrounds—in physics, chemistry, biology, engineering, and medicine—and fostered a culture in which interdisciplinary research thrived.

Getting students to recognize the value of multidisciplinary teamwork meant getting them into the lab to begin focusing on their research projects—and Lewis was in favor of making that all happen earlier rather than later in a student's career. Rigorous coursework that taught the fundamentals—thermodynamics, kinetics, statistical mechanics—was crucial to a budding scientist's professional future. But excellence in research and a wealth of experimental know-how was even more important, in Lewis' view. So he helped institute a policy that ensured that graduate students began conducting research in their first year of study.

In fact, the qualifying examination for PhD candidacy, a detailed exam through which professors judge whether a student is well suited to work toward a doctoral degree, was taken in the second year of graduate studies. The timing of that exam, which then like now, required second-year students to demonstrate detailed knowledge of the concepts and hands-on laboratory experience, is early even compared to other universities today.

In 1957, I didn't know a whole lot of Berkeley's chemistry history. But I knew enough to recognize that I was at a world-class research institution and that the decisions I made would shape my future. One of those decisions, a personal one, was that a teaching assistantship stipend of about $150 a month was enough income to start a family. So Judy and I got married just before Labor Day that year and moved into an apartment on Delaware Street in Berkeley. We enjoyed a three-day honeymoon at Lake Tahoe during the holiday weekend.

(1957)   *(2012)*

*For more than 55 years, Judy and I have been
blessed with a beautiful marriage.*

Sounds luxurious, doesn't it? Actually, out of necessity, we economized everywhere. My stipend covered our $65 per month rent bill and from what remained, we paid for food. And that was about it! Our transportation costs were kept to a minimum by living within walking distance of campus. We weren't exactly living the high life in those days.

We talked amongst ourselves about ways to make a little money here or there and even bounced an idea back and forth between some other Hungarian students of opening a Hungarian restaurant. There weren't any at the time in Berkeley—and it seemed like a solid idea. But as we talked it over, we soon recognized that starting such a venture would require a capital investment—and none of us had any capital to invest! What's more, running such an operation would require investing a huge amount of time—a commodity that full-time students typically find to be in short supply. That put an abrupt end to the restaurant idea.

Meanwhile, I had to figure out how to earn some income during the summer months because my teaching assistantship was in effect only during the nine-month school year. So I landed a job at a local company, Stauffer Chemicals, synthesizing and studying the properties of polycarbonates, a class of polymers that had only recently been discovered. That job tied me over financially until the next school year began and gave me a taste of what is was like to work in industry at a very early stage in my career.

Judy and I also took various odd jobs during our university years. At one point, I tried my hand as a door-to-door salesman selling men's electric shavers. I was horrible at it. I don't think I made even one sale! And Judy worked for a while as an *au pair* in Berkeley and at Lake Tahoe and also as a cashier at Kress's five & dime, an inexpensive variety store. She also worked on an assembly line at a paper box factory in Oakland.

It was during the summer of 1957 that I began looking in earnest for a professor with whom to work. Anyone who was ever involved in pursuing a graduate degree in the sciences, especially a doctorate, understands the structure and culture of graduate level research. But for readers who are not familiar with this form of education, let me explain.

A small portion of a student's years in graduate school, typically at the very beginning, is almost identical to undergraduate education; you attend lectures, take notes, study a textbook, and take tests.

In contrast, the large majority of a student's years working toward a PhD in the sciences is devoted to a research project. (A smaller portion is spent teaching—often undergraduate laboratory courses.) The research project is typically part of a larger research program managed by a professor (sometimes more than one professor). The student becomes a member of the professor's research group and is considered a research assistant. It's a day-in and day-out kind of science occupation—and the research group, depending on the personalities of its members, is in many way like an extended family. Science grad students generally spend thousands of hours per year, literally, working with or in close proximity to other members of their research group. The experience often forges lifelong relationships between students and between students and professors.

Now it was time for me to decide whose research group I wanted to join. There were two fields in chemistry that really caught my attention at that time; catalysis and polymers. Both of them fascinated me, but I must admit, I knew very little about either subject. As it happens, no one at Berkeley was working directly on those subjects in those days. But Richard Powell, a Berkeley chemistry professor and inorganic kineticist (someone who studies the rates at which metals and their compounds undergo chemical reactions), agreed to take me on as a graduate student.

My project was to investigate small platinum particles of the type that were used as catalysts in industry, for example in petroleum refining. I'll explain more about these materials later, but for now let it suffice to say that unlike chemical reactants (also called reagents) that are consumed in a reaction, catalysts are not consumed. Their presence alone makes many types of reactions proceed quickly (or proceed altogether) and do so with a smaller input of energy than would be required in their absence. There are plenty of examples of catalytic reactions, including ones that help convert crude oil to automotive fuels and ones that rid automobile exhaust of pollutants.

To make the most of precious metals like platinum, manufacturers who make catalysts for those types of applications typically prepare the metals as microscopic particles that are attached or embedded— "supported" is the common scientific term—on ceramic materials that are essentially inert or unreactive. No sense in making large clumps of platinum because chemical reactions only occur on their surfaces—their exteriors. All the platinum inside a particle's ultrathin outer layer is hidden away and thus wasted.

The trouble is, with extended use inside a chemical reactor, catalyst particles change size—they usually grow by aggregating and fusing to one another and eventually forming large particles. They also become fouled, "poisoned," or deactivated, as scientists say, through accumulation of some kind of muck on their surfaces. When I was a graduate student, little was known about the way metal catalyst particles become deactivated. Professor Powell proposed I study that subject by probing the microscopic bits of metal with X-rays.

In particular, the idea was for me to use a method that was fairly new at that time known as small-angle X-ray scattering. That technique seemed like a good choice because, in principle, it could provide information about the sizes and shapes of particles in the nanometer-range, that is, billionths of a meter across. Catalyst samples would be made available to me by one of Professor Powell's former students who was a manager at that time at Standard Oil of California in nearby Richmond. So I decided to join Professor Powell's research group. At the time, he only had two students. Nonetheless, he was a first-rate mentor, a jack of all trades, a man with a deep knowledge in all areas of chemistry who was brimming with enthusiasm. I learned much from him.

One thing I had not counted on learning was that he was one of those academics, who at that time, opposed the influx of federal funding into scientific research. People who shared his view—and there were many of them in academia then—feared that the considerable inpouring of government dollars during the *Sputnik* and Cold War era would compromise freedom of research in universities.

I cannot speak for Professor Powell and I certainly understand the need for an academic scientist to be free to choose his or her research path. But it is clear that federal support for scientific research in the U.S. helped make the United States the world leader it is in scientific and technical innovation. That outcome may not have been so easy to predict back then. It's far easier to look back several decades and map the rapid growth of federal agencies charged with stimulating and supporting scientific research as well as the growth in funding they allocated for those projects.

There are many ways to trace the rise of government support of scientific research in the U.S. But let me begin by telling you about one event that changed things overnight: *Sputnik I.* On October 4th 1957 (October 5th in the eastern hemisphere), the Soviet news agency announced the successful launch of the world's first man made Earth-orbiting satellite.

At just 184 pounds and 22 inches in diameter, the design of the aluminum sphere which raced around the globe every 96 minutes was not exactly awe inspiring. Actually it was perfectly ordinary. But the fact that the Soviet Union had beaten the United States into space came as colossal shock—a piercing wakeup call—to U.S. policy makers. Overnight the U.S.S.R. was able to legitimately and publicly claim technological superiority over the U.S.—and that's exactly what it did.

And less than one month later, on November 3rd, when the Soviet Union sent a dog into orbit aboard the seven-times-larger *Sputnik II*, the perceived technological gap between the U.S. and U.S.S.R. crushed the image of America as a powerhouse of scientific know-how. The blow was felt by Americans everywhere especially on Capitol Hill. Accused of doing little to ensure global supremacy for U.S. science, the government in general—and the Eisenhower administration in particular—moved quickly to regain America's confidence. After debating several proposals for the creation of an agency charged with planning and directing aeronautical and space activities, the National Aeronautics and Space Administration (NASA) was created in the summer of 1958. NASA's budget, like that of the overall U.S. investment in basic and applied scientific research, began to shoot up exponentially. Within a few short years, the space agency commanded an annual budget of some $35 billion.

At the same time, the U.S. government was pouring money into other agencies responsible for encouraging education and research in the sciences and mathematics. These funds radically altered the nature of scientific research on U.S. college campuses. For example, the National Science Foundation was established in 1950 during the presidency of Harry S. Truman with an initial budget of roughly $150,000. In 1952, with $3.5 million available for research support, NSF funded 28 scientific research proposals.

But when fears of scientific and technical inferiority gripped the nation following the 1957 *Sputnik* launches, Congress sprang into action. NSF's appropriation was increased in one year by nearly $100 million to $134 million in 1959. By 1960, NSF funded 2000 research projects. By the 1980s, with a budget exceeding $1 billion, the agency supported some 12,000 projects.

U.S. spending on energy research also increased dramatically during that period—and as it did, the structure of the government body overseeing that investment changed with it. After World War II and the Manhattan Project era, Congress established the Atomic Energy Commission to maintain civilian government control over atomic (or nuclear) research and development. During the early part of the Cold War, much of AEC's focus was on nuclear weapons, but in the 1950s its emphasis shifted to overseeing nuclear power production.

Everything changed again in response to the oil and energy crises of the 1970s. With Americans suddenly aware of how dependent their way of life was on petroleum supplies and how uncertain those supplies were, the U.S. government formulated the Energy Reorganization Act of 1974.

That piece of legislation put an end to AEC and established in its stead the Nuclear Regulatory Commission, which as its name suggests, was formed to regulate the nuclear power industry, and the Energy Research and Development Administration. ERDA was established, among other reasons, to facilitate and manage basic and applied scientific research programs aimed at developing and improving fossil-fuel, solar, alternative energy, and energy-conservation technologies. ERDA gave way in 1977 to the U.S. Department of Energy, which was established to coordinate a federal-level comprehensive national energy plan.

Throughout their various incarnations, the energy agencies of the U.S. government have invested heavily in university and other science-institution research programs. The goal has generally been to promote scientific and technological discoveries through research projects that would not or could not be conducted or financed in the private sector. Much the same can be said about federal support for science outside of the energy sector. You can get the abridged version of the story, simply by looking at the numbers.

The overall level of federal R&D spending skyrocketed from roughly $8 billion in 1950 to something on the order of $150 billion in 2010. During the same time frame, U.S. spending on non-defense-related research climbed from about $1 billion to some $60 billion. The largest allocation increases at the end of that period were awarded to the National Institutes of Health, the Department of Energy, and the National Science Foundation.

Without a doubt, the level of federal funding provided for scientific research at American universities in the 1950s was a real game changer in the world of academic scientific investigation. And in the 50+ years since that time, federal research dollars have continued to stimulate cutting edge discovery and innovation making U.S. science second to none.

It just so happens, that Richard Powell, who I decided to work with to pursue my doctorate, was one of those academics who did not take federal research grants. He wasn't alone. Plenty of faculty members at UC Berkeley in the 1950s were leery of excessive federal influence in academia. It was the age of McCarthyism, so-named for Joseph McCarthy, the Republican senator from Wisconsin who came to symbolize the anti-communist hysteria that purveyed the American psyche in those days. Often with little evidence to support allegations, Americans in every walk of life were being accused of disloyalty, anti-American behavior, or treason and relieved from their posts or otherwise fired.

The effect of suddenly and publicly suspecting someone who only yesterday seemed perfectly forthright and trustworthy was felt acutely in Berkeley. In 1949, University of California faculty learned of a new California law requiring them to take an Oath of Loyalty to the United States and swear that they are not members of the Communist party. Incensed by the oath's infringement on academic freedom, some sixty UC Berkeley faculty members quickly banded together to oppose the law.

The atmosphere turned bitterly contentious. As mistrust of suspected Communists grew in some circles, mistrust of government grew in others. On Capitol Hill at that time, the House Un-American Activities Committee opened an inquiry into the alleged infiltration of Communists into Berkeley's Radiation Laboratory. And in California, the University of California Regents, the body that governs the U. C. system, expressed concerns that the university had also been infiltrated by Communists.

Faculty were soon told to take (or sign) the oath or be fired. 150 of them refused to sign. After a series of committee hearings in 1950, 31 faculty members were fired. Eventually, in 1952, California's Supreme Court ordered that they be reinstated. Nonetheless the situation remained tense. As reports from that period show, some 50 people who were offered U. C. faculty positions declined the offers at least in part due to the Loyalty Oath controversy. Concerns over government meddling in academia, especially in California, remained high for years.

As a result of Professor's Powell's concerns and his stance against taking federal research grants, he had little money to spend on laboratory equipment. He also had no funds for other research expenses or with which to support graduate students. That, in turn, meant I had to work as a teaching assistant (usually teaching lab courses or small lecture sections) every semester in order to be paid by the university. Many professors supported their students financially, which freed them from the chore of teaching undergraduates giving them more time for their own lab work.

The situation wasn't ideal, but it was what it was and I had no choice but to make the best of it. I ended up borrowing equipment from Professor David Templeton, a well-known X-ray crystallographer who had served on the Manhattan Project and had been recruited to U.C. Berkeley by Glenn Seaborg. Unlike my professor, Templeton had research funds and used them to build instruments with which he made important contributions to X-ray crystallography. Much the same was true of other Berkeley physical chemists, such as George C. Pimentel, who creatively advanced the field of gas-phase molecular spectroscopy, and Kenneth Pitzer, a prominent figure in molecular thermodynamics who served as the Atomic Energy Commission's research director from 1949 to 1951.

The key point to note is that although faulty members had differences in ideology regarding federal financial support for research, nonetheless, a spirit of camaraderie suffused the halls of science at Berkeley in those days. Professors, post docs, students, and staff were generally all happy to help one another. Scientists there shared a collegial let's-work-together kind of attitude—and I benefited from that mindset.

So with no equipment to work with in my professor's lab, I borrowed equipment from Professor Templeton's lab and set about building a small-angle X-ray scattering machine. It didn't take long to complete. Soon I was using the apparatus to learn how the sizes and shapes of these platinum catalyst particles responded to high gas pressures and temperatures. I passed my qualifying examination in 1958 and then committed myself to working long hours in the laboratory.

But first, to celebrate my success, Judy and I went to Montgomery Ward and bought our first television set. I must have spent three days sitting in our apartment in front of that little T.V. marveling at the way Popeye the Sailor Man was able to make all of his troubles disappear just by eating a can of spinach. If only things worked that way in real life! Of course, in addition to the joy of owning our first T.V. (we paid it off slowly; a couple of dollars per month), we also had a brand new and interesting way of learning English and learning about American life. At the same time, I worked diligently on my PhD studies. In a total of three years, I completed the work for my PhD degree, which was granted to me in January 1960.

Along the way, I learned a few important lessons about experimental physical chemistry, which are as relevant now as they were in 1960. To begin with, I told myself, don't shy away from new laboratory methods simply because the techniques are unfamiliar or the instruments seemingly complicated. New methods and tools may help uncover new fundamental information. Don't be afraid of them; master them.

Next; making discoveries in science isn't just a matter of conducting experiments and recording data faithfully. It often comes down to designing the *right* experiment. Frequently, that means simplifying most experimental conditions, reducing the number of variables, and minimizing irrelevant parameters, so that the phenomenon in which you are most interested stands out clearly. This "minimalistic" or "reductionist" approach to experimentation would eventually come to characterize much of the work I did in the surface chemistry of catalysts.

The real frontiers of research in the physical sciences, including surface chemistry and catalysis—the subjects of my doctoral thesis— lie at the atomic or molecular level. Understanding the structures of molecules that adhere to (are adsorbed on) solid surfaces and the tendencies of chemical bonds in those molecules to undergo change are where cutting edge research has long been focused. For decades, scientists wanted to understand how molecules adsorb on surfaces, move around, undergo chemical reactions, and leave (desorb) from surfaces.

Although the fields of gas-phase spectroscopy and X-ray crystallography had already made great strides in revealing molecular-level information, by the late 1950s, science had not yet developed the tools that could provide information about surfaces at that level of detail. Still, I remained impressed by the relevance of this research to so many processes in daily life—not just the ones taking place inside industrial chemical reactors—and I decided to return to this area if ever the means became available to investigate surfaces on the molecular level.

Towards the end of 1958, it was clear that I would be finishing my PhD work in a year's time, which meant that I would be looking for employment—and considering whether to move away—before Judy would be able to complete her undergraduate studies in chemistry. She decided then to take a leave of absence from school and went to work at Standard Oil of California in Richmond. We were glad to have the additional income.

Meanwhile, my parents were living in Chile at that time. My sister, aunt, and uncle had been there for a while and my parents wanted to join them—mostly to be near my sister. So they ended up surrendering their home in Budapest to an agent of the Hungarian AVO in exchange for passports. What a trade!

We wanted to visit with them, but that was tricky on several accounts. First of all it was difficult for them to obtain U.S. visas because Judy and I were not U.S. citizens at that time. And travel was terribly expensive for us. So we wrote to Pan American Airways and explained our situation—refugee students, no money, family abroad—and they very kindly provided us with inexpensive airline tickets.

I remember stopping in Panama City on the way there and back and touring the Panama Canal and a few other sights before continuing on our way.

Judy's parents and brother, on the other hand, had come to the U.S. and settled in Berkeley in 1958. They had given up their apartment in Budapest, made their way to Vienna, and eventually qualified for special refugee status, which put them on an expedited track for emigrating to the U.S. As long as I mention Judy's family, let me jump forward and point out that Judy's brother, Andrew (Andy) Kaldor, was another family member who went into chemistry. He did undergraduate work at Berkeley and went on to Cornell University, where he earned a PhD in physical chemistry. Eventually he ended up at Exxon (later ExxonMobil) where he spent many years as a manager in research and development.

Back in 1958, near the end of my tenure as a graduate student, I began interviewing with several technology companies in search of employment. I had been teaching nonstop for three years and had little interest in pursuing an academic position. So I talked with representatives from DuPont, IBM, Kellogg, Shell, and Chevron. Times must have been good; they all offered me positions. After some deliberation, I finally accepted a research position with IBM in Poughkeepsie, New York.

Truth be told, IBM offered me the highest starting salary, $10,500 per year. But there were more than dollar signs in my eyes. Somehow, I felt that by going to a company with relatively few chemists, I would have a chance to shine—and my skills would be valued. I also considered that I would have opportunities to learn new areas of physics, electrical engineering, and computer science, topics about which I knew next to nothing.

In January 1960, after completing all of my PhD work, we walked around Berkeley, saying sad goodbyes to all of our friends and looking, for what we thought would be a final time, to the campus and the area near our new home that had been brand new to us just a few years earlier. We hardly expected to return. Then we flew to New York and continued on by train to Poughkeepsie.

IBM had generously offered to pay our relocation expenses, but we had nothing to take with us, except for a 1951 Chevrolet. After Judy had been working for a while, we bought our first car. We were in for more than a few surprises.

We really knew nothing about automobiles, least of all inexpensive used ones. Living in a densely populated metropolitan area in Europe, our families, like those of most of our friends at that time, did not drive. Mainly we traveled by train.

We learned the hard way that our first car, which we were so proud to own, didn't tolerate being driven at speeds above 40 miles per hour. We found out that driving faster caused one of the ill-shaped piston rods to break. So wherever we went, we got there by driving at 40 mph or slower.

Another curious feature of our first car was that it lacked a heater. So when I proudly asked IBM to ship the car to Poughkeepsie for us in January, we ended up freezing during the first several months whenever we drove it. What's more, coming from Hungary by way of California, I knew nothing about winter driving conditions. I was O.K. in chemistry, I had a handle on physics and mathematics, but I certainly didn't know anything about snow tires!

One last "fond" memory about the car is that it put us in a bit of a financial pickle. When I received my first paycheck, I was shocked to see how small it was. It turned out that IBM had added the relocation expenses and the cost for shipping the car to my annual salary, but had taxed the entire sum and deducted it from that first paycheck. As a result we did not have enough money left from that check to pay our house rent. With great embarrassment, I went to the IBM personnel office and explained what had happened and asked for a loan. Fortunately, the people there helped us out of that uncomfortable situation.

As long as I mention the house, let me add that we had rented this small house because it was close to IBM and Vassar College. Vassar is a well-established liberal arts school in New York's Hudson Valley. Judy attended there for two years to complete her chemistry degree. Then as now, Vassar is quite selective and tuition is high. So in order to help defray the cost of attending a private college, Judy applied for and received a grant from the American Chemical Society's Petroleum Research Fund. The grant supported Judy for two summers as she worked with an advisor as an undergraduate researcher. I am extremely proud to note that our two oldest granddaughters, Stephanie and Clara, are attending Vassar College. As Stephanie was completing her freshman year, her grandmother was there celebrating her 50th college reunion.

Not long after I started at IBM, it seemed to us as though we had accumulated a great deal of money, since I was now making a small fortune relative to the tiny stipend I received as a graduate student. I bought my first suit. We went to restaurants. We drove the car to New York City to go to the opera and the theatre. And by the fall of 1960, I am proud to say that we bought a brand new car.

When I began working at IBM, the company was in the final stages of completing its new research headquarters in Yorktown Heights, New York—about 40 miles south of Poughkeepsie. By the time summer arrived, the entire Poughkeepsie laboratory relocated to Yorktown Heights. Not only was it exciting to be working for a major American technology company such as IBM—and to be working alongside numerous bright and creative thinkers, it was truly special to be doing cutting-edge science at IBM's brand new T. J. Watson Research Center.

The imposing crescent-shaped edifice in Yorktown Heights was the creation of Eero Saarinen, a highly regarded Finnish-American architect. Saarinen's sweeping, arching industrial design concepts made him a sought after architect by American corporations such as General Motors, CBS, and others in addition to IBM.

Regardless of personal history, someone who experiences milestone events such as getting married, completing doctoral studies, and starting work at a just-completed state-of-the-art research facility—all in the span of a few short years, surely must take stock of where they are and where they are headed in life.

We did all that. But we also took stock of where we came from. Just a few years earlier we were new immigrants in a new land—immersed in a new language and culture and unsure of what the future held for us. And now we were moving up steadily and preparing to embark on the next chapter of our lives.

CHAPTER 5

Work and Life in Industry

My IBM Years: 1960-1964

*Technology and basic science. Innovation and invention.
U.S. high tech industry was an amazing place for a
young scientist to be in the 1960s. Experimentation
and discovery was clearly my calling.*

THE EARLY 1960S was an especially fascinating time to be working at a leading technology company in the United States. Inventions, discoveries, and technological developments of all sorts were sweeping through the halls of centers of innovation, such as IBM and Bell Labs. The needs and demands of the U.S. military during the Second World War spurred a flurry of research activity in communication technology, high-altitude aviation and rocketry, computing, and other areas. After the war ended, that research boom continued at a furious pace.

A number of technology areas that, in principle, could have developed along separate paths came to be intertwined in mutually beneficial ways. Progress in computing, for example, depended upon complex and cumbersome machines—early computers—which were built from large numbers of vacuum tubes, resistors, switches, and other electronic components. As computing developed, the field of electronics developed too—especially blossoming with the inventions of the transistor and integrated circuit.

Transistors were a new type of electronic amplifier and switch that, for household applications such as radios, and specialized applications such as computers, could replace the much larger vacuum tubes used at that time. Vacuum tubes, which somewhat resemble conventional incandescent light bulbs, do their job by means of an electric current derived from a glowing metal filament. Transistors, in contrast, perform similar jobs by means of the electronic properties of a crystal of a semiconductor such as silicon or germanium. These new transistors were relatively small and many of them could be fashioned into a single miniature integrated circuit, which promised complex yet small devices.

The invention of these devices signaled tremendous potential advantages for manufacturing and commercialization due to a collection of useful properties—only one of which was the transistors' small size relative to vacuum tubes. Examples of other useful properties, some of which took a while to put to commercial use, include lower weight, operating voltages, and power consumption, higher reliability, longer service life, and the potential for automated and low-cost manufacturing.

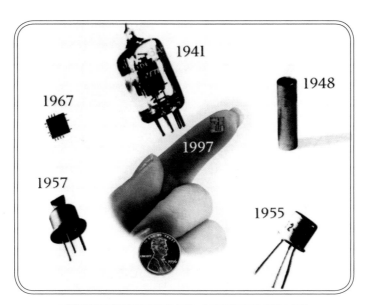

And the Shrink Goes On....

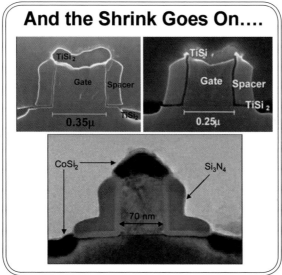

From radio and vacuum tubes of yesteryear to modern transistors, electronic amplifier devices have shrunk steadily over six decades. The smaller the overall size, the greater the importance of surfaces per electronic device. The drive towards ever greater numbers of tinier and tinier circuits has led to lightning fast computer processing speeds and today's powerful handheld devices.

At much the same time, the space race, in which the U.S. was deeply engaged, called for exploiting the latest advances in electronics to develop new types of space-worthy instruments. The drive to advance space technology also spurred development of vacuum equipment to evaluate how such instruments and other types of machinery would fare in the vacuum of space. It also led to new types of laboratory instruments that depended on the advances in vacuum technology.

The intersection of all of these areas—semiconductors, electronics, transistors, vacuum technology, and novel lab tools—is where I landed when I arrived at IBM. Many of these topics were rather new to me; they weren't the sorts of subjects with which young chemists usually have much experience. That made things really interesting for me. Not only that, but most of my coworkers at that time were physicists and electrical engineers. That also made things interesting for me—but in a different way. Not only was I able to learn from them about all those topics and others, like solid state physics and computer programming, but I was able to make unique technical contributions and help solve problems by drawing on my background in chemistry.

An opportunity to put that chemistry to good use came my way shortly after I arrived at IBM. At that time, IBM was interested in using the semiconductors, cadmium selenide and cadmium sulfide, to build inexpensive calculators. Exposing those materials to light generates a flow of electric current through the semiconductors. IBM was exploring the possibility of using a flashing light to create pulses of electric current in those semiconductors so as to represent the 0s and 1s of binary code to run the calculators.

The problem was that after a short period of use, the power level dropped precipitously. My job was to figure out why—and what could be done to prevent it. So I studied the way these semiconductors behaved upon exposure to light and whether that behavior changed in the presence or absence of air. By using the controlled atmosphere of a vacuum system, I was able to control those variables and figure out what was happening.

It turned out that the light was triggering a reaction—a photochemical reaction—between oxygen (in air) and the semiconductor surface. That process rather quickly formed a thin coating of cadmium oxide on top of the semiconductor which impeded the events that generated electric current. That change in the chemical composition of the crystal's surface was irreversible. Once it formed, it was stuck there. I reasoned that it should be possible to find a suitable material to encapsulate the semiconductor and protect it from oxygen exposure.

All in all, this project went well. I did experiments that pinpointed what was causing the trouble, explained to IBM that the cadmium compounds could not be used as is, and proposed a way around the problem. It was decided however, that encapsulating the materials would drive up the cost dramatically and so the project was abandoned by IBM. Yet this research experience and related ones, coupled with the atmosphere of innovation and scientific exploration at IBM, helped forge the path I would end up following in research for decades—the path through surface science.

The molecular-level details of the way in which the surface of the cadmium compounds changed upon exposure to light and oxygen were unknown to me and to the world of science during my early days at IBM. I was only able to construct a general picture of those processes. But I was very curious about the elementary details. Some of the earliest laboratory instruments that could probe those chemical processes—at that level of detail—were just beginning to be developed, and I was keen to get my hands on them.

Several factors come into play simultaneously. For starters, the key aspect of the semiconductor crystals I was studying is their surfaces. Advances in electronics were just starting to take off that time, and that meant that electronic components were shrinking. As the little bits of semiconductor crystals at the heart of the new transistors and other circuit elements of the day were getting smaller, the ratio of their surfaces to their bulk was getting larger. Nowadays, transistors have dimensions that measure in the low nanometer—billionths of a meter—range. But even back then when transistors were much larger, as the size of the crystal shrunk, the importance of the composition and chemical nature of its surface grew.

Here's a way to picture the surface-to-bulk ratio. Imagine tearing off the paper from a gift-wrapped house and wadding it up into a tight ball. That ball is tiny compared with the size of the house. But if the gift wrapped object is a dollar bill, which effectively has no bulk, the wrapping paper (the surface area) and the dollar bill (the bulk) are nearly the same size.

But even if they weren't similar in size, a thin surface—just a handful of molecular layers—could dominate many of the bulk object's properties. In the case of the house, or even a dense piece of lumber, for example, a thin coat of paint can prevent the wood from absorbing moisture and rotting; repel mold and mildew; and cause the wood to absorb or reflect sunlight, depending on paint color.

The same is true of extremely thin anti-reflection and anti-glare coatings on camera lenses; optical coatings on "polaroid" sunglasses; various types of anti-fog, germicide, and ultraviolet-blocking coatings; and metal films on plastic food packaging, which form a gas diffusion barrier and keep foods fresh. Similarly, thin films at the surface of contact lenses and medical implants make the difference between products and devices that are comfortable and helpful to lens wearers and patients and ones that are rejected by the body. In all of these examples, and in many more, an ultrathin skin—perhaps a few molecular layers in total—can dominate the properties of a comparatively thick bulk object.

It stands to reason that in order to study the surface of any kind of object—a little chunk of semiconductor for electronics or metal catalyst particles like the ones I examined in graduate school—the first order of business would be ensuring that the sample is clean and pure. In that way, one could compare the properties of the sample before and after its surface has been altered, coated, or gunked up. In many cases, ensuring sample cleanliness means handling the sample in vacuum—that is, in the absence of air or other gases.

Commercial vacuum technology had been around for decades by the time I arrived at IBM. It was central to the thriving vacuum tube industry, for example. But as a result of the space race, vacuum techniques, tools, and hardware improved markedly in the 1960s. That made it fairly easy to buy laboratory vacuum systems that included a vacuum chamber—an engineered metal can in which experiments were to be performed, pumps capable of pulling out (nearly) all of the air to reach extremely low pressures (or high vacuum), and other equipment such as vacuum gauges (specialized pressure gauges) for monitoring the vacuum condition.

Because our bodies are well suited to functioning and thriving under standard air pressure conditions (like the ones in the room where you're reading this book), we tend to give little notice to the air all around us. But there's a tremendous amount of it under normal atmospheric conditions. How much is "tremendous?" On a typical day at sea level, every cubic centimeter of air contains more than ten billion billion molecules, mostly nitrogen and oxygen. With every breath we draw, we pull trillions of billions of molecules into our lungs.

With all that gas around flying to and fro, a freshly cleaned sample of semiconductor, metal, glass, etc…, immediately becomes coated with layers of gas molecules as they impinge on the sample surface. Cleaning the sample under those conditions is futile. It's like wiping clean a mirror in a steamy sauna; in seconds it gets fogged over again. That's one of the main reasons for conducting many types of surface science experiments in a vacuum chamber. Conditions known as ultrahigh vacuum (UHV), in which the gas pressure is reduced below one trillionth of standard atmospheric pressure, can be achieved routinely with off the shelf equipment. Under UHV conditions, a sample can be cleaned, often by heating it to drive off gas molecules that adhere (adsorb) to its surface, and it can stay clean for hours—long enough to study its native surface. It can also be exposed or treated inside the vacuum chamber with select gases to induce surface chemical reactions.

Another reason for depending on vacuum equipment to study the surfaces of solids is connected to the techniques used for probing surfaces. That brings us to those early instruments for studying molecular-level surface processes, the instruments I said earlier I was eager to get my hands on.

In the early 1960s, instruments for conducting low energy electron diffraction or LEED experiments were just coming to market. The discovery of LEED dates back to the 1920s and is in itself a fascinating story. But it wasn't until key advances were made in electronics and vacuum technology—around 1960—that LEED instruments were starting to be available for purchase.

The essence of the LEED experiment is that by firing a beam of low energy electrons at a crystal surface, the electrons scatter from the surface in an orderly fashion. That orderliness reflects the atomic or molecular structure of the topmost layers of the crystal— its surface. And that's the ticket to deriving molecular-scale surface information.

More on that topic in a moment. For now, let me just emphasize that in order for electrons to travel from the hot metal filament in an electron gun to the sample—and to bounce off the sample surface in an orderly way and fly to an electron detector, the medium through which they travel must be a darn good vacuum. Otherwise, the electrons will collide with air molecules and never reach their destination. In fact, these collisions will occur so frequently even under low vacuum conditions, that the situation is almost like trying to play baseball under water. The ball would never reach the batter. And even if the batter did manage to hit the ball, it would never leave the batter's box.

Improvements in vacuum technology took care of those problems. It made it possible for electrons (and later on ions and beams of atoms and molecules) to travel long distances (centimeters) in a vacuum chamber without colliding with air molecules and abruptly changing directions. That long "mean free path," as it's known, would be essential to a host of future vacuum technologies.

In the case of LEED, Clinton Davisson and Lester Germer experimented in the 1920s by firing beams of electrons at metal targets inside glass vacuum tubes. As Germer himself explained to me (and as he described in the July 1964 issue of *Physics Today*), the pair was conducting these experiments to substantiate their position in a patent suit regarding vacuum tube technology rights between Western Electric Company, which later became Bell Telephone Laboratories, and General Electric.

Through a confluence of unlikely events (a hot vacuum tube broke) the conditions were just right for a nickel target to crystallize and scatter the electron beam in the orderly manner I mentioned earlier. Initially, this peculiar type of scattering made little sense to the Bell Labs researchers. But eventually the scientists learned of the recently formulated electron wave theory of Sorbonne University physicist Louis de Broglie. Davisson and Germer reasoned that what they were observing wasn't garden variety electron scattering but rather diffraction, a special "flaring out" process that de Broglie had recently predicted should happen with electrons, but which until that time was only known to occur with light.

For de Broglie's revolutionary theory describing the wave nature of electrons, he was honored with the 1929 Nobel Prize in Physics. Davisson and George P. Thomson of London University, who made the earliest high energy electron diffraction measurements, were recognized with the 1937 Nobel Prize in Physics for their experimental discovery of electron diffraction.

With this advance it was possible to fire a beam of electrons at a crystal surface—and by analyzing the pattern of diffracted beams, deduce the geometric arrangement of atoms in the surface layers. The process is analogous in many ways to X-ray diffraction from the interior of crystal lattices. To conceptualize this process, it may be helpful to draw an analogy between atoms in a crystal and eggs in an egg carton. Like eggs in a carton, which are arranged neatly with fixed distances between every pair, so too atoms in a crystal are arranged regularly with repeating distance intervals. Now picture a large table completely covered with one layer of filled egg cartons. That image represents the topmost layer of a crystal surface.

In one manufacturer's carton, the distance between neighboring eggs in one row (call it the x direction) may be the same as the distance between adjacent rows (the y direction) —say two inches each. In another carton, those distances may differ a little. If we cover the table with a second layer of egg cartons on top of the first, and then a third and fourth and so on, we can slowly build up a 3-dimensional egg crystal.

It may be that the spacing between the layers (the z distance) is different than the other two distances. If we then cut this imaginary crystal in several ways—say at "90°" angles and at oblique angles along various diagonals—each time exposing a new surface (pretend the eggs don't break!), then we will certainly end up with surfaces that feature distinct geometric patterns and spacings between neighbors. It's those regularly repeating distances that determine the pattern of diffracted beams. In the case of LEED, electrons can neither penetrate deep into the crystal nor escape from deep within the bulk of the crystal lattice. And so the diffraction patterns are characteristic of just the top few atomic layers. That's key!

A few more points; just as slicing a crystal along various lattice directions can expose various surfaces characterized by distinct atomic spacings, which can be investigated via LEED, so too a host of other processes can lead to noteworthy surface geometries. For example, exposing a crystal to a reactive gas such as oxygen may alter the geometry of the surface layer. Or the gas may adsorb in an orderly pattern that differs from the underlying surface structure. Heating the crystal may alter (or restore) its geometry. And in some crystals, the topmost layer or two may spontaneously rearrange because the surface "prefers" an energetically relaxed geometry.

That gives you a flavor of the kinds of surface information that can be investigated via LEED. For more than 30 years, however, the potential to exploit this discovery "sat" as it were, in a quiet corner of science. Then around 1960, the vacuum technology and electronics needed to take LEED forward were developed. One of the key advances was the design of an apparatus to accelerate the diffracted beams onto a fluorescent screen. That setup yielded a bright display of the spot pattern, which could be viewed, photographed, and studied to obtain quantitative surface information.

I became increasingly interested in surface science and decided to focus on that field. I made arrangements to visit Bell Labs in New Jersey to meet with the handful of researchers working on LEED and came away even more enthusiastic about this developing field and its cutting edge research tools.

And so it was that I persuaded my manager to request that IBM purchase a LEED instrument so that we could begin our own surface experiments. IBM agreed. To my knowledge, we purchased the first commercial LEED unit made by Varian, a California instrument maker. The machine arrived and I happily began setting it up and started getting to work.

No sooner had I begun getting my new LEED system up and running than I was paid a visit by my manager. He complimented me on my research success and promptly informed me that I was to be promoted to group leader of semiconductor crystal growth. I certainly was honored and thankful that my work had been recognized and that IBM wanted to promote me, but I explained that I really was interested in surface chemistry and related phenomena. I tried to make a case that I could be an asset to IBM by working in those fields.

He listened to me patiently and then completely ignoring everything I said, he repeated, "Well, congratulations, you're all set to become the new group leader of semiconductor compound crystal growth." It was clear that I wasn't being given a choice. IBM had decided that it was time for me to move on to areas of greater technological importance and possibly greater financial importance to the company. I should have been proud about the promotion offer—and in a way I was proud. But I was unhappy about the prospect of making a scientific *about face*. I really wanted to continue working in surface science.

So that's when I decided it was time to leave IBM and find a university faculty position, which would give me research independence. I applied to various universities, and in the fall of 1963, I set off for interviews at CalTech, U.C. Berkeley, and U.C. Davis.

Unforgettably, the day I went to CalTech, President Kennedy was assassinated. In the middle of my interview, one of the faculty members rushed in and announced the terrible news from Dallas. Obviously, the interview ended abruptly. I went back to finish the process two weeks later.

Fortunately, I received offers from everywhere I interviewed, including two offers from Berkeley—one for a position in chemical engineering and the other in chemistry. With these offers in hand, I knew I wanted to go back to Berkeley, my alma mater. I also knew that in terms of subject matter, the offer from the chemistry department was the right one for me.

Judy and I had other reasons to go back to Berkeley. My parents had moved to the United States in 1962, after I became a citizen, and at that time they were living in Berkeley, as was my sister who was married by then. Judy's parents and brother were also living in Berkeley then—and so going back to Berkeley meant living close to our family again.

It was at that time that Judy became pregnant with our first child, our daughter Nicole. It was such an exciting time for us as an expecting young couple. We were so happy and our friends were so happy for us. We both worked at IBM—Judy as a computer programmer and me as a scientific researcher. And so when Judy's due date drew near, our friends from work and from our neighborhood made three baby showers for her. Judy recalls being so loaded down with lovely gifts that for a while, our house looked like a baby store.

Nicole was born June 16, 1964. The birth of this beautiful little baby was a blessing and a wonderful experience in our lives. She was so tiny and seemed so fragile, I was almost afraid to hold her. But in reality she was healthy and strong and I was just a nervous new father. The arrival of this precious addition to our growing family marked the beginning of a wonderful new phase in our lives. Just two weeks after Nicole was born, we boarded a plane headed for San Francisco and returned to Berkeley.

CHAPTER 6

Faculty Life in Berkeley

From Assistant Professor to Member of the National Academy of Sciences (1964-1979)
Surface Chemistry at Low Pressures

A career in academic research affords young professors freedom to choose topics of investigation, motivation to discover new phenomena, and an opportunity to make an impact in science. Yet sometimes it requires defending unexpected conclusions that contradict conventional wisdom.

IN JUNE OF 1964, Judy and I returned to Berkeley with our two-week old daughter, Nicole, found a suitable apartment near campus, and settled in. I was given an office in one of the chemistry buildings, but there was no laboratory space available at that time. The new chemistry building, which would be called Hildebrand Hall, was being built when I arrived, but it wasn't ready yet.

There was lab space available in the new Inorganic Materials Building in the Lawrence Berkeley National Laboratory (LBNL), just a just a few minutes' drive from the UC Berkeley campus. At that time, the Laboratory's focus was starting to undergo a shift from nuclear sciences to materials sciences, and the Lab director, Leo Brewer (the expert in heat-resistant inorganic materials I mentioned earlier) kindly helped me out.

Leo offered me lab space, funding, and a position as a faculty scientist at LBNL. I gratefully accepted the offer and used the funds to begin equipping my laboratory. The first thing I bought was an ultrahigh vacuum system and a low-energy electron diffraction instrument from Varian. To my knowledge, that instrument, which arrived in the fall of 1964, was just the second such system sold by Varian. (The first system was the one I used at IBM, which remained there when I returned to Berkeley.)

As you can imagine, I was raring to get started in my new lab. But just as I was starting to find my way during my first few months as an assistant professor, life on the Berkeley campus took a crazy turn. It was there and then that thousands of students staged demonstrations in support of the Berkeley Free Speech Movement. These demonstrations, which were led somewhat informally by political activists, the most famous of whom was Mario Savio, were intended to force the university leadership to acknowledge students' right to free speech and to permit students to openly promote on-campus political activities. Such activities were officially banned at that time.

The Free Speech demonstrations turned chaotic when one of the group's activists was arrested by campus police in October of that year for refusing to show an I.D. card. Accounts of the episode relate how, in a drawn out and tense standoff between demonstrators and police, a few thousand students surrounded the squad car in which the arrested activist sat. Eventually, he was released.

Not long thereafter, thousands of students staged a multiday demonstration in front of Sproul Hall. Photos and film footage of the events, including Mario Savio's impassioned speech, are readily available on the internet; they give a strong sense of what the intensity of that era was like. The events culminated in mass arrests, which triggered even more demonstrations and widespread civil disobedience across campus.

I was astounded when the demonstrations turned violent. Police were forcibly trying to maintain order, and students were physically preventing people from walking freely on campus. It happened to me.

One day as I walking from the chemistry buildings towards Sather Gate, I found demonstrators blocking the entrance to the gate; they weren't letting me or anyone else pass through. I was enraged. What kind of nonsense was that? So I marched right up to them and tried to get by. Immediately, I was grabbed by several rather husky demonstrators and held for a few minutes, but eventually they let me go and I continued on my way.

On the way back, I deliberately wanted to walk right through the demonstrators' line again, but some security guards warned me that that would just be asking for trouble. So I capitulated and walked around the long way. In some ways, all these experiences reminded me of the senseless actions of the Hungarian police that I had witnessed years earlier.

Although I remember these events clearly, and although they came to be part of the fabric of U.C. Berkeley life, I was only involved marginally. Most of these scenes played out on the west side of campus, where the humanities departments were located. The east side, where the science and engineering departments were primarily located, was much quieter—and for the most part, I was busy trying to get my footing as an assistant professor.

Shortly after I took up my faculty position, two graduate students decided to work toward their PhD's in my laboratory. As I explained earlier, graduate students and post-doctoral associates (students who continue working with a professor after obtaining a PhD) form the core of a university science research group. They work with the professor conducting investigations. Or as they develop skills in experimentation, they may work somewhat independently and consult with the professor for guidance. Either way, a professor needs to attract students to build a research group, so having two students sign on with me was quite a big deal.

One of the students was older than me. He was a military pilot who was sent to Berkeley to earn a PhD. The other student came from Rice University. I also attracted a postdoctoral researcher by the name of Stig Hagstrom from Sweden. Stig had been a student of Kai Siegbahn, who was a co-winner of the 1981 Nobel Prize in Physics for his work in electron spectroscopy. So when Stig joined my group, he already had a background in vacuum technology and was especially interested in X-ray photoelectron spectroscopy. That technique was eventually developed into a powerful tool for analyzing the chemical composition of surfaces—meaning it can be used to distinguish and identify the specific chemical elements in the topmost atomic layers of a material. This type of spectroscopy came to be quite important in my work.

So together with my three-person research team, I began to study surfaces. I recognized how important it was to understand the properties of catalyst surfaces from my PhD work and electronic component surfaces from my days at IBM. It would be quite a few years until I more fully appreciated just how ubiquitous and important surface properties are across so many more and such wide ranging fields. Let me give you a sense of what I mean.

As a young professor, I was thinking about the physical properties of crystal surfaces—their structure, composition, arrangement of atoms, interaction with microscopic particles and light—and how those physical properties affect surface chemistry, especially in connection with catalysis. As time marched on and scientists and engineers learned to exert fine control over structure, composition, and other physical properties, they developed catalyst technology with tailored chemical properties. Examples include highly efficient catalytic chemical plants that produce the fuels that power the world. And environmental clean-up technology such as automobile catalytic converters, which today convert engine exhaust to pristine and pollution-free gas.

In the mid-1960s I wasn't thinking about biological surfaces and interfaces. But those too, are everywhere and incredibly important. Consider that the external surface of every leaf is "charged" with the task of collecting sunlight to stimulate the photosynthesis processes on which plants thrive. And that the internal surfaces of bone are designed to stimulate vascularization and adequate transport of blood and nutrients to growing tissues. Consider also that life-saving coronary stents and all body-implanted devices, such as bone and joint replacements, must be designed and treated for surface biocompatibility. If internal body surfaces are "unhappy" with an implantable device's surface, the body will reject the foreign part. Clearly, the field that I selected as my research area turned out to be quite a good one—and even more important than I realized initially.

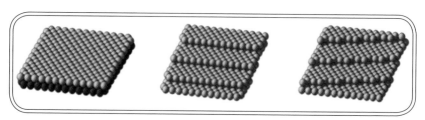

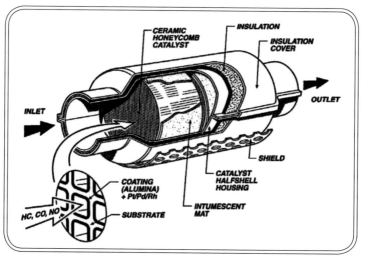

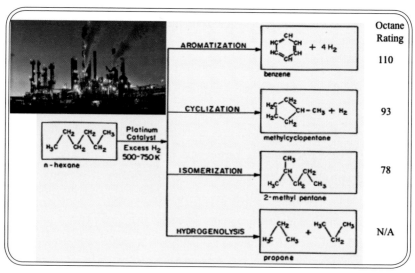

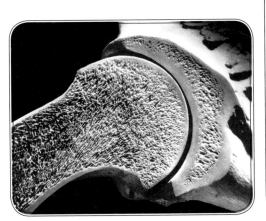

Surfaces are ubiquitous and important in so many ways. Fine control over surface structure and other properties of microscopic catalyst crystals (three gold-colored models) enables automobile catalytic converters and modern chemical plants to function with extreme efficiency. Biological surfaces and interfaces are central to healthy bone development and photosynthesis in leaves (above) and can dictate the longevity of medical implants such as coronary stents and bone joints (below).

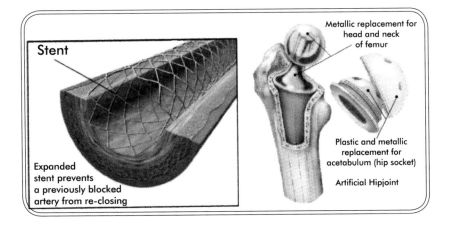

Stent

Expanded stent prevents a previously blocked artery from re-closing

Metallic replacement for head and neck of femur

Plastic and metallic replacement for acetabulum (hip socket)

Artificial Hipjoint

Selecting "good" topics for investigation is vitally important to an academic researcher's career. At first blush, it may sound strange to someone who has never been involved with scientific research to hear that research is carried out competitively; but it is.

Researchers generally have to compete for limited funds by submitting research proposals that are evaluated critically. If the reviewers judge that a proposal is scientifically unsound or feel that the researcher is unlikely to succeed, perhaps because the goals are too difficult to achieve, the proposal will be rejected and someone else will win the grant. And after funds are procured, determining how to get the best return on the research investment is crucial.

A research group leader needs to be successful scientifically in order to secure promotions and tenure; to publish articles in respected peer-reviewed journals; and to attract talented students and postdocs, who make a research group thrive. Being successful scientifically means, in part, being the first person to make significant discoveries in a research field.

Most discoveries advance science incrementally not enormously; that is, they add some new information or uncover a previously unnoticed piece of a puzzle that deepens understanding about some phenomenon. Very rarely does someone discover something that truly turns science upside down. Sometimes the media or even historical accounts play up a certain discovery as singlehandedly revolutionizing some area of science. But that description of the scientific process is not very realistic and it certainly does not reflect typical developments in science. Rather, science and the careers of scientists generally advance step-by-step through carefully planned studies and thoroughly analyzed research results.

I decided to focus on the surface chemistry of platinum and other catalytic metals. There were a number of reasons for this choice. To begin with, there were very few researchers studying metal surfaces at that time, so there was room in the field for a young investigator to make meaningful contributions. In contrast, semiconductor surfaces were being investigated by advanced teams of researchers at places like Bell Labs and IBM. It was unrealistic to think that a new university investigator with a very small research team would be able to compete successfully with those larger and better equipped teams. In addition, I had some background knowledge in catalytic metals from my graduate school days in Richard Powell's laboratory, and all importantly, the topic still fascinated me.

In many ways, platinum is considered the grandfather or archetypical metal catalyst. The term "catalysis," which is derived from the Greek word meaning to loosen or untie was coined in the 1830s by Jöns Jacob Berzelius, a Swedish scientist and one of the world's foremost chemists in his day. Berzelius first used the term in one of his reviews of physical science—annual reports he prepared for the Swedish Academy of Sciences—to describe a collection of phenomena that for nearly 20 years prior had been attracting considerable attention across Europe. Chief among those phenomena was platinum's special ability to spontaneously ignite hydrogen in air.

Imagine the thinking of Berzelius and other leading chemists of nearly 200 hundred years ago as they tried to understand this property of platinum. Unlike the oxygen (in air) and hydrogen that were converted to water and thus consumed in the chemical reaction that generated the flame, platinum was not consumed. Yet the metal's presence was a prerequisite for combustion. Without it, nothing happened.

Nowadays, we understand that catalysts facilitate and speed up chemical reactions by causing reactants to quickly get into the right configuration or state that immediately precedes some critical step in a chemical reaction. That key step, for example, could be formation of a preliminary bond between two molecules. Even after taking the initial step that begins to unite the molecules, the reactants may still need to do a little molecular choreography—rotate this way, bend and swivel that way—to form the final chemical product. And perhaps the reactant molecules can be coerced into taking that first step, for example by cranking up the heat to force them to come together. But in the presence of a catalyst, that molecular union comes about with much less effort and energy because the catalyst makes the conditions for getting together just right.

In some ways it's like a host who throws a fun party and smoothly introduces one of his bachelor friends to a lovely bachelorette. It's possible that one day the young man and woman may have eventually met on their own and hit it off. But through the workings of their mutual friend, the whole process took place much more quickly and effortlessly.

Long into the 20th century, many of the details of these catalyzed molecular unions or other processes were unknown. But in the 1800s they were already a hot research topic. Even before Berzelius came up with the term "catalysis" in the 1830s, he and others were already trying to understand the phenomenon. In the early 1820s, he wrote in his annual reports about the investigations of Johann Wolfgang Döbereiner, a German chemist who in 1823 discovered that by directing a stream of hydrogen at a platinum gauze, the metal became white hot and spontaneously ignited the gas. That observation was the basis for what eventually was called the *Döbereinersche Feuerzeug*, a kind of lamp that was used as a household lighter and manufactured by the tens of thousands. So-called safety matches had not yet been invented and here was a convenient way to light candles and make a fire for cooking and heating.

Döbereiner's work quickly caught the attention of the British chemist Humphrey Davy, who had been involved with similar work and had discovered related properties of platinum some years earlier. Davy had been investigating the flammability of gases such as methane, which posed serious explosion hazards to coal miners and led to many mining accidents in the early 1800s. Miners obviously needed light to work underground. But the exposed flames of the candles or oil lamps they carried touched off deadly explosions when flammable mine gases wafted near the flames.

Davy's work led him to design a widely used coal miner's safety lamp, in which the flame was shielded by a fine metal mesh. Methane and air could readily pass through the mesh to the flame and burn under control. But the flame would not leap through the metal screen and cause an explosion. Not only did the lamp safely provide light in an explosive atmosphere, but by observing changes either in the glow of a platinum wire or of the flame shielded within the wire mesh, it also served as a detector for flammable gases.

Humphrey Davy's younger cousin, Edmund Davy, also studied platinum. He found that the powdered form of the metal turned white hot in the presence of alcohol vapor and stayed that way until the alcohol was consumed. Nearly the same results were observed when he exposed powdered platinum to coal gas, a common fuel gas mixture at that time containing methane, hydrogen, and carbon monoxide.

Long after these early catalysis researchers pondered the way platinum mediates these chemical transformations, the valuable metal became an industrial workhorse catalyst. Platinum was eventually used on a commercial scale to oxidize ammonia, hydrogenate (add hydrogen to) organic molecules, and convert low-performance and low-octane transportation fuels into high-octane fuels. The puzzle pieces that have since been fit together to explain how this metal works its catalytic magic have been discovered slowly. And in the 1960s when I was starting to build a research group, few of the puzzle pieces had been uncovered.

Surface reconstruction of platinum and other metals

To get started studying platinum's surface properties, I purchased a special kind of a platinum specimen known as a single-crystal rod. Single crystals are similar to high-quality jewelry gemstones in that all of the atoms more or less line up with crystalline order throughout the whole piece of material. That high level of atomic-scale perfection is not found in metals purchased from hardware or tool suppliers. Single-crystal metal rods, which sometimes are the size of a short pencil or a finger, are painstakingly prepared by specialty materials companies—and they can be quite expensive!

Think back to the stacks-of-egg-cartons analogy I made to describe the orderliness of a crystal lattice. I explained that if you sliced that imaginary egg crystal in various ways—for example at "90°" angles and at oblique angles along various diagonals—then you would end up exposing surfaces that feature distinct geometric patterns and spacings between neighboring eggs. I learned how to do that with platinum and was able to cut the rod in a way such that the freshly exposed surface was the so-called (100) face. Crystallographers use that three-digit notation to name the whole range of faces that can conceivably be exposed by slicing a crystal at all sorts of angles.

My students and I also learned how to prepare and clean the crystal so that it was suitable for surface experiments. Typically, some combination of sputtering the crystal face with fast moving ions—a kind of atomic-scale sandblasting that dislodges molecular "dirt" from the surface—coupled with high-temperature heat treatments— all done inside a vacuum chamber—leaves the exposed crystal face pristine and ready for experiments. Cleaning routines that work well for one material may not quite do the job for another. So a little trial-and-error experimentation is needed in order to learn how best to prepare each crystal for surface studies.

After all the preparation was completed, we examined the crystal with our new low-energy-electron-diffraction instrument to see if we could observe a LEED pattern. Just to be clear, if you stick a doorknob or a piece of an automobile bumper in front of a LEED instrument (also called LEED optics), you won't see a LEED pattern of bright spots because those pieces of metal aren't crystals—their atoms are not aligned in an infinite number of regimented rows and columns.

When electrons fired from a LEED optics strike these ordinary metal specimens and ricochet back toward the instrument's phosphor screen, they strike everywhere, not just in select spots as they do when the scatter from the surface of a crystal. For that reason, electrons scattered from the surface of an off-the-shelf piece of amorphous piece of metal cannot generate a neat LEED pattern. You can imagine then, just how eager we were to see what would happen when we fired electrons at our specially-cut platinum crystal.

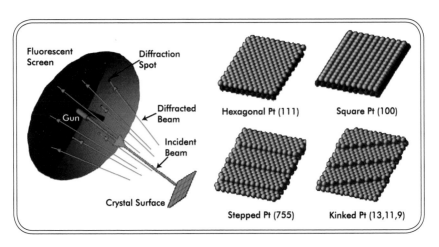

Low energy electron diffraction (LEED) can reveal details about the arrangement of atoms and molecules at solid surfaces and can distinguish among the geometric patterns they form (ball model examples). The technique involves firing a beam of electrons from an electron gun at a crystal surface and studying the pattern of spots formed on a fluorescent screen as the electrons ricochet from the crystal.

Sure enough, by looking through the window (or viewport) that was deliberately built into the vacuum chamber so as to provide a clear view of the LEED instrument's phosphor screen, I saw a neat pattern of bright spots. Simply seeing the spots was already exciting! That meant that my vacuum equipment was doing its job of reducing the gas pressure inside the chamber where the crystal was held to near one trillionth of standard atmospheric pressure. It also meant that our new LEED instrument was working. And quite importantly, it meant that after cutting this piece of platinum, which induces a lot of atomic disorder in the exposed faces, we had succeeded in restoring that order.

Strangely, instead of seeing the relatively simple square-shaped pattern that we were expecting to see—the one that theoretically should have arisen from electrons bouncing off platinum's (100) face, I saw a very busy pattern marked by all kinds of "extra" spots. This face of platinum was definitely crystalline but it wasn't the native (100) face.

It took some while before we understood what this unexpected pattern signified. But soon we recognized that the relative positions of the atoms at the surface of the crystal were different from their positions deep inside the crystal—in the bulk. Compared with the atomic pattern of a bulk (100) plane, the (100) plane at the exposed face—the crystal surface—is reconstructed. The arrangement of atoms in the reconstructed face is "busier" than its analog many layers below the surface and that's what was causing the "busy" LEED pattern.

In 1965, I published a paper reporting the discovery of reconstruction of platinum surfaces in *Physical Review Letters*, a well-known journal. The article was received enthusiastically by scientists working with semiconductors. Researchers in that community were already familiar with surface reconstruction from studies conducted at Brown University and Bell Laboratories on semiconductors such as silicon, germanium, and gallium arsenide. To that group, it seemed reasonable that metals and other covalently-bonded materials (materials in which all atoms "share" electrons evenly) would also reconstruct.

Surface scientists who studied metals, however, were skeptical of our findings. Their experiences showed that metals were notoriously full of impurities as a result of the processes used in metal refining. Metals rarely exist in pure form in nature. Rather, they tend to be found in the form of oxide and sulfide minerals and often mixed with other metals. The chemical methods used for extracting, separating, and purifying metals invariably leave behind traces of those processes. 99.9% pure, for example, may sound like an incredibly high level of purity. But 0.1% concentration of impurities can easily change a material's properties. The metals experts figured that impurities had something to do with our reconstruction results.

In those days, there was a group of surface scientists who studied metals using the related methods of field emission microscopy and field ion microscopy. Those techniques were used to study the positions of atoms in an incredibly sharp metal needle, often tungsten, by applying a high voltage to the needle, which causes electrons or ions to fly toward a phosphor screen. One of the most prominent members of that group was Robert Gomer of the University of Chicago.

I reported on our platinum reconstruction work at a scientific conference where Gomer was present. He wasn't convinced that the unexpected LEED pattern we observed was due to surface reconstruction of platinum. I remember he said something to me like "Young man, tungsten is the only metal that can be prepared without surface impurities because it alone can be heated to high enough temperature to boil off all the impurities."

Tungsten has an incredibly high melting temperature—the highest of any chemical element: 3410 °C or 6170 °F. That's why it's often used as light bulb filaments, which need to glow white hot for very long periods. The thinking among the metals experts was that unlike tungsten, platinum could not be heated high enough to scrupulously clean it of impurities, and these trace impurities somehow had a hand in causing the unexpected LEED results.

Naturally, I was disappointed—even somewhat worried—that an established expert in surface science didn't accept my explanation of our experimental results. So I needed to come up with a way to prove that surface reconstruction was a property of platinum and not of impurities. But in those early days of what we now call "modern surface science," there really weren't any generally applicable instrumental techniques for determining the identity of atoms and molecules adsorbed on surfaces.

It seemed reasonable to check whether platinum's neighbors in the Periodic Table of Elements, iridium and gold, behaved as platinum did. Due to the way iridium (element number 77), platinum (78), and gold (79) are refined, it is unlikely that the same impurities would be found in all three elements. So we also examined crystals of those elements and sure enough, we found the same type of surface reconstruction. Those results were certainly encouraging but they were not solid proof.

Fortunately, there were a number of researchers who were interested in our results and some of them, including Thor Rhodin and Paul W. Palmberg of Cornell University, were interested enough to repeat our studies. Rhodin was a professor in Cornell's department of applied and engineering physics, and at that time, Palmberg was a research associate there. Both scientists led careers that focused heavily on developing instrumental methods for surface analysis.

In the Cornell version of the experiments, rather than cutting a single crystal of platinum as we did, a piece of the metal was heated high enough inside a vacuum chamber to cause atoms to evaporate (sublime) from the sample material and settle on a cooler surface. If the surface onto which sublimed atoms settle has a structure that's compatible with the material being sublimed, and if the temperature and other conditions are just right, then the deposited atoms can coalesce and slowly build up a thin *crystalline* film of material B on top of material A. If conditions aren't just right, this deposition process can proceed anyways, but the deposited film won't be a thin crystal; it will be amorphous, like the film of filament metal (often tungsten) that darkens the inside of incandescent light bulbs after prolonged use.

In the Cornell study, platinum was deposited onto the surface of a magnesium oxide crystal. And just as in our case, LEED analysis showed that the surface of the thin platinum crystal was reconstructed. That was quite significant because whereas our pencil-sized piece of platinum may have contained residual impurities left over from processing conditions, the Cornell "crystal" was an ultrathin film of platinum that was created via sublimation in the pristine environment of high vacuum. That thin crystal was certainly impurity-free.

So there we had it—proof, confirmation that surface reconstruction was an inherent property of the material. The whole experience was educational and reaffirmed the workings of the "scientific method" that elementary students learn about today. It works like so: scientists discover something new and report the findings; other scientists evaluate the claims—some are convinced, others are skeptical; independent researchers repeat the experiment; if the results match (preferably in more than one lab), the original discovery is validated.

Our findings were reproduced independently and I was starting to establish a solid record as a principal investigator.

Life seemed to be moving quickly in those days. Our small family grew in 1966 when Judy gave birth to our second child, a wonderful, healthy little boy, whom we named John. It was time to buy a house—and even though money was tight (I took a substantial pay cut when I left IBM for academia), things were going well at the university and I was certain that I would soon be granted tenure. With tenure, which came in 1967, I was able to enjoy the benefits of job security and focus on my research. So we looked around and found a lovely house—and after securing two mortgages and a loan from the seller, we bought that house and moved in. More than five decades later, I can say we have lived happily in that house ever since.

In that house we raised our two children, Nicole and John. Nicole became a Ph.D. chemist, and is married to Paul Alivisatos, a highly respected U. C. Berkeley chemistry professor. They have two daughters, Stephanie and Clara. Nicole serves as coordinating editor of *Nano Letters,* a prestigious American Chemical Society journal and Paul serves as coeditor-in-chief. John is a business executive in the software industry, and his wife, Hilary, works for Harvard Business School. They have two children, Benjamin and Diana. We are very fortunate that all our loved ones live in northern California, within an hour's drive of Berkeley. We have been truly blessed with a beautiful family.

Developments in Surface Analysis Tools Determining Structure, Composition, and Oxidation State

Things were also moving quickly in science at that time. New types of surface analysis tools were being developed—and with these tools, researchers were able to start identifying the chemical species on solid surfaces. Quite importantly, these instruments were also starting to be commercialized widely in the late 1960s, which meant that they weren't solely available to instrument innovators, but rather to the community of surface science researchers.

Two of the most important techniques developed in that period were X-ray photoelectron spectroscopy and Auger electron spectroscopy. As their names indicate, both methods depend on detecting electrons emitted by samples. In each case, those electrons carry chemical information about the material from which they originated. Now, electrons are found throughout the entire volume of any piece of material—they don't just reside at the surface. But due to the way electrons interact with solids, and the difficulty they have in sailing through layers and layers of atoms, the electrons analyzed by these two spectroscopy methods originate strictly from the topmost atomic layers of a solid. That's crucial. Because the electrons come only from a material's surface layers, the chemical information they carry comes strictly from those layers too. That's very different from most chemical analysis methods.

The two electron spectroscopy techniques share a lot in common. In both cases, an atomic-scale jolt kicks electrons out of the specimen causing them to fly through the empty space of vacuum to an electron analyzer that measures their kinetic energies. The chemical information they carry is encoded in that energy.

Much of the development of X-ray photoelectron spectroscopy, which is usually abbreviated XPS or ESCA for Electron Spectroscopy for Chemical Analysis, is credited to Kai Siegbahn, the 1981 Swedish Nobel laureate whose student, Stig Hagstrom was my post doc at that time.

The essence of the method is that firing X-rays at a material causes electrons deep in the cores of atoms—meaning electrons that reside near an atom's nucleus—to be ejected. Those core electrons are bound to the nucleus by an energy that is characteristic of each type of nucleus—that is, each chemical element—and of the specific electron core level from which it is ejected. In addition, details of an atom's chemical environment, for example subtle differences in the nature of carbon bound up in methane (CH_4), benzene (C_6H_6), or carbon monoxide (CO), also leave their characteristic mark on the core electrons' binding strengths.

Instrument developments soon made it fairly simple to determine those all-important binding energies, which provide unique signatures for each element and their chemical (or oxidation) states. Because X-ray photons have a fixed energy that exceeds the minimum energy needed to kick out these electrons, and because the energy for each type of X-ray source is a known quantity, an electron's binding energy was soon easily determined from its kinetic energy, which is the parameter actually measured by the instrument.

Auger electron spectroscopy (pronounced oh-JAY with a soft J) is quite similar to XPS, with a twist. The technique is named for Pierre Victor Auger, a French physicist who studied electron emission but wasn't actually was the person who discovered the process that carries his name. Credit for that discovery actually goes to the Austrian physicist and chemist Lise Meitner. In any case, similar to XPS, Auger electron emission is also triggered by an atomic-scale jolt.

Firing X-rays or energetic electrons at a solid can kick out a core-level electron, as just explained, but in the Auger process, the ejection or excitation event is followed by an internal electron shake-up.

Here's what happens: the ejected electron leaves behind a core-level vacancy—a hole—deep inside an atom. The energetic instability caused by the sudden appearance of that vacancy causes a neighboring core-level electron, one from an energy level that's higher than the ejected electron, to "fall" in energy level and fill the vacancy. But that sudden change in energy levels further propagates the shake-up process.

The difference in energy between the "falling" electron's original and final energy levels is imparted to yet another (a third and sometimes even a fourth) core level electron. That third electron, suddenly energized by the whole process, is promptly ejected from the atom. That electron, which is known as an Auger electron, also carries element-specific information encoded in its kinetic energy.

The details and chronology of the instrument developments that made these measurements routine could be the subject of a whole book on the history of surface analysis equipment. No need to go into all of that here. But one important thing to note is that a number of instrument innovations made by General Electric and by Paul W. Palmberg of Cornell University (mentioned earlier) together with some of his colleagues at the University of Minnesota led to a commercially available LEED instrument that could be used for LEED and Auger analysis. Instruments that coupled those originally separate analytical techniques soon became available commercially from Physical Electronics, a company of which Palmberg was a cofounder. Collectively, these instrumental developments majorly changed the face of surface science by broadening the discipline to include analysis of surface chemical reactions.

So the time was right for me to start moving from studies of platinum's surface structure to catalytic reactions mediated by platinum. By this point, we had learned to cut and prepare crystals so as to expose a number of crystal faces—not just the (100) face. And so we began adsorbing organic molecules on various faces of platinum. LEED analysis revealed a collection of neat spot patterns, which indicated that the molecules were settling on platinum's faces in an orderly fashion. But how were they settling there? What kind of bonds were the organic molecules forming as they came in contact with platinum? If for example, platinum was exposed to ethane (H_3C—CH_3), did that molecule alter its C—C bonds and C—H bonds to form fresh bonds with platinum? Did the molecule change or break apart when it landed on the metal?

Unfortunately we were limited at that time in our ability to interpret those well-formed LEED patterns because the theory underlying the electron diffraction process was not yet well enough developed. Eventually, over the course of the next 10 years or so, theoreticians were able to draw upon early LEED measurements to help construct and test a theoretical framework for interpreting complex LEED patterns. Near the end of the 1970s it became possible to use LEED results (which were recorded with a film camera) to determine rather accurately the positions of atoms, relative distances between atoms, as well as the lengths and angles of surface chemical bonds.

By 1968, the surface structure work that my group and others were doing had generated enough interest that I was able to organize and host a conference on that subject at Berkeley. "The Structure and Chemistry of Solid Surfaces" drew a few hundred people who were actively researching or wanted to learn more about semiconductor and metal surfaces and their emerging applications in microelectronics, coatings, and sensors. The meeting and my role as its organizer thrust my name onto the national surface science scene for the first time and helped me begin establishing a reputation as an up and coming "player" in the field.

The conference was memorable not just because of the impressive attendance and positive feedback it generated but also on account of some embarrassing technical problems.

The first speaker at the symposium was the famous Erwin Müller, a fastidious German physicist who was a professor at Penn State and the inventor of the field-ion and field-emission microscopes. Müller, who in 1955, together with his graduate student Kanwar Bahadur, was the first person in the world to see the image of an individual atom, had a reputation of being not such an easy going guy.

Before the days of Power Point presentations projected from laptop computers, speakers came to deliver their talks armed with a slide-projector cassette loaded with their presentation slides. The speaker would stand near the front of the room and the projector operator would sit somewhere in the middle of the room, waiting for the speaker's cue to advance the slides.

"May I have the first slide?" Müller called. BAM! The projector bulb popped. No problem. I had a spare ready and replaced the blown-out bulb. Three precious minutes later we started again. "First slide, please." BAM! The second bulb exploded. Eventually I found a bulb that worked properly and the symposium finally got underway but I was a little worried that Professor Müller may have thought I had a hand in trying to derail his lecture.

Sabbatical in Europe

As a young professor at Berkeley, I regularly taught the undergraduate physical chemistry lecture and laboratory courses. The university had a policy that an instructor who accumulated 3½ years of teaching credit was entitled to take a six-month sabbatical and after seven years, a one year sabbatical. In 1969, I was eligible for my first six-month sabbatical, so we made plans to spend the time in Europe. It turned out to be a wonderful experience in which we made new friends, established new contacts in European labs, and saw beautiful sights.

I chose to go to England to the University of Cambridge. Jack Linnett, a professor of physical chemistry there, was kind enough to host me and arrange a visiting fellowship at Emmanuel College, which is one of the colleges at Cambridge. At that time, I also won a Guggenheim Fellowship, which was a prestigious honor that came with a stipend. So in the spring of 1969, Judy and I—together with our two young children—arrived in England. I quickly came to enjoy the local customs—like coffee served in the chemistry department in the mornings and tea time in the afternoons. The British ways were elegant and heartwarming. But that's about all they warmed. Because it was springtime, the heat had been turned off in the laboratories— they said they didn't need heat in the spring—and I remember clearly that it was chilly all day long.

As a visiting scientist, I moved freely between the departments of chemistry and physics and soon met everyone who had any interest in surface science and surface chemistry. That's where I met David Tabor, a physics professor who was a pioneering researcher in tribology, a field of immense technological significance that's dedicated to investigating friction, wear, and lubrication.

Another noteworthy acquaintance I made then was that of John Pendry, a young student in the Cavendish Laboratory. Pendry went on to become a leader in theoretical solid state physics and the key figure responsible for developing the theoretical basis of low energy electron diffraction. His work in that area is what made it possible for us—a few years later—to quantitatively interpret our LEED data.

"Interpret" means deducing the atomic positions, relative distances, and bond angles that characterize the ordered surface layers that we were probing.

The time spent at Cambridge was wonderful. During that time, Jack Linnnett and I developed a close friendship that lasted until he passed away in 1979. Also during that period, because I was free from teaching duties, I had time to begin writing my first book, "Principles of Surface Chemistry," which was published in 1972 by Prentice Hall.

Writing a science textbook, or a "monograph," which is similar to a textbook but not intended for new students, gives the author a detailed perspective on the way the subject of the book—surface science, in this case—developed. In the usual course of events in science, researchers plan and carry out experiments, analyze the results, and report the findings in the form of peer-reviewed scholarly journal articles ("papers") if the findings are novel and interesting enough to share with other scientists. Each of these papers covers a very narrowly defined subject—generally a single study.

After a substantial number of journal papers have been written on a subject, the editors of a journal may call upon one of the field's experts to review the entire body of literature in that field (or in that sub-specialty) in a comprehensive review article. Review articles can be excellent reference sources for scientists—including graduate students—who are trying to "come up to speed" on a given subject. After several years of research have firmly demonstrated that scientists are genuinely interested in the subject of these review articles, a publisher may ask one of the field's experts to write a monograph. These books are broader types of reviews—ones that for example, may include more history, applications, and future research directions than typically included in a journal review article. Monographs are generally intended for experienced scientists.

Finally, after the information contained in the reviews is distilled and refined, and the field has developed enough that it sports a unique collection of important findings and well-known case studies, then the time has come to write a textbook for university students. This whole process by which scientific knowledge is uncovered, debated, developed, refined, condensed, and finally taught in a matter-of-fact way to young students fascinates me. Some of the concepts we teach freshman chemistry students today were the subjects of cutting edge research just a couple of decades ago. Of course, science, like other subjects, has fads that come and go. For a while some subjects are popular and researched intensively and then dropped for lack of interest. But that too is part of the scientific method. The important discoveries and concepts persist and eventually show up on the pages of classic textbooks.

After spending a little while in Cambridge, we traveled by boat from Harwich on England's North Sea coast to Esbjerg in southwest Denmark. That was our first time visiting Denmark. There we spent time with Haldor Topsøe, founder of the large Danish catalyst company that bears his name. We also visited Göteborg (Gothenburg), Sweden, and I had a chance to spend some time at Chalmers University of Technology with physicists Ingvar Marklund, Stig Andersson, and Bengt Kasemo, all of whom studied surface science and were knowledgeable in electron diffraction techniques. From there we went on to Germany, where I met Gerhard Ertl, who at that time was a professor of physical chemistry at the Technical University of Hannover. Ertl's areas of interest overlapped closely with mine and we had a chance to discuss our LEED investigations of surface molecular layers.

Coming back to Europe as a tenured professor at the University of California 12 years after fleeing Europe as a refugee was an incredible experience. I was full of hope and excited for the future; I was thankful that my life had moved on the way it did; and I was proud of my professional accomplishments. All in all it was wonderful to be back visiting Europe. But one dark episode has stayed with me vividly all these years.

I had arranged to visit a senior scientist in Göttingen, a small town with a famous university close to the former East German border. I regret that after all these years his name slips my mind, but nonetheless, he kindly invited me and offered to pick us up at our hotel and drive us to his apartment for dinner. The plans sounded just fine. Now, I had never met the gentleman face-to-face before. So when he pulled up to the hotel door—after dark—and introduced himself to us while sitting in his car, I couldn't see him very clearly and had no way of recognizing him. We got into the car and he drove us to an apartment building where he parked in a dimly lit underground garage.

Suddenly I was overcome with anxiety. What if the driver wasn't really who he said he was? What if we were being kidnapped by an East German agent? What if, what if…I didn't know what to think and I didn't want to scare my family. But for the next few minutes, which seemed like an hour, I was on edge, my heart racing. We stepped into an elevator, went up a few floors, got off, walked down the hall, and he showed us into a lovely apartment—and in the end, we enjoyed a delightful dinner.

It just goes to show how traumatic experiences can leave an indelible mark on our psyches. A dozen years after living through fearful experiences in Hungary, I couldn't shake off the silly fears that were crowding my mind during that short car ride.

As long as I am mentioning trips to Europe, I'll add that in 1974 we returned to Europe for another brief sabbatical stay. This time we started at Bristol University, where I was honored as "Visiting Unilever Professor" and was asked to deliver a series of lectures. Since Nicole and John were 10 and 8 years old at that time and relatively enthusiastic travelers, we decided to use the opportunity to visit other European countries that we had not yet visited.

We went to Trieste, Italy, and to Paris, where we visited with Jacques Oudar and Christian Minot, catalysis specialists at the University of Pierre and Marie Curie. We also went to Strasbourg in eastern France, where we visited with Prof. Goldstaub and his students, three of whom later worked with me as post docs.

We also spent time in Lyon and then went by train to Madrid, where we visited Juan F. Garcia de la Banda, director of Madrid's catalysis institute. Finally, we ended our whirlwind tour in Zurich. It was a wonderful trip and marvelous experience for me to see surface science and catalysis unfold in all its diversity—internationally.

Molecular Beam—Surface Scattering Experiments

It was during this time that I decided to use molecular beams to study surface chemistry. At a simple level, a molecular beam resembles a stream of water steadily flowing from a garden hose—or perhaps a "feed and weed" chemical mixture as it is sprayed on suburban lawns by yard and tree-care specialists. The stream issues forth—or shoots out—from a nozzle and can be carefully directed at a target. Similarly, a beam of gas molecules (including solids and liquids that have been vaporized) can be conveyed to a target by way of a directed stream, instead of floating aimlessly in a nebulous cloud.

To get to the heart of what goes on at the surfaces of catalytic metals—that is, to begin to understand these processes at the molecular level, I began planning molecular beam experiments. The idea was to use vacuum techniques to direct a stream of molecules at a single crystal surface that had been studied and characterized ahead of time, to try to determine what happens to the molecules. Do they form bonds to the metal and stay put? Do they bounce off (scatter from) the surface and leave unchanged—meaning, without undergoing chemical reactions? Or do they make and break chemical bonds and undergo molecular rearrangements?

If the beam is composed of reactive molecules, as opposed to ones that are chemically dull (inert), one would expect that monitoring them after they have interacted with a catalytic metal, for example with a mass spectrometer positioned close to the surface, would reveal interesting changes in the molecules that make up the beam. Perhaps the molecules would land on the surface, break apart and find new molecular partners with whom to bond, and eventually leave (desorb) from the surface. Those were the kinds of molecular scenarios we aimed to explore in our new experiments.

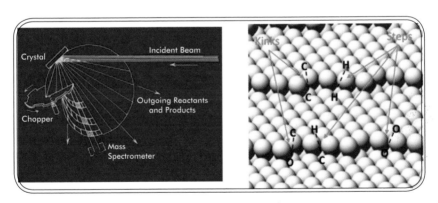

In a molecular beam-surface scattering experiment (left), a beam of molecules impinges upon a crystal surface and scatters. A mass spectrometer can be used to determine the identities of molecular species that bounce back, and a device known as a chopper can measure molecules' velocities. A laboratory setup such as this one can be used to follow surface dissociation of molecules. This technique has revealed that dissociation of H-H, C-C, C=O, and O=O bonds preferentially occurs at atomic steps and kinks—not along terraces which tend be less active.

As I mentioned earlier, Berkeley had a long tradition of building novel scientific instruments that dates back to the days of G. N. Lewis and the machine shop he established for the chemistry department. That "we can design and build it" attitude was reinforced by many Berkeley scientists such as E. O. Lawrence, Glenn Seaborg, and others. In particular, Dudley Herschbach and Yuan Lee had already built instruments at Berkeley for producing crossed molecular beams. (Together with John C. Polanyi, they were honored with the 1986 Nobel Prize in Chemistry for their studies of elementary chemical dynamics.) So the idea of me designing and building my own molecular beam apparatus at Berkeley to study chemical reactions on surfaces—and the rates of those reactions, which is known as chemical kinetics, was not overly daunting; it was a doable project.

We started off with simple scattering experiments—using inert gases such as helium and argon—to gain some experience in using the instrument and making measurements. Then we moved on to a more complex process known as hydrogen–deuterium exchange. Deuterium is an isotope of hydrogen—that is, chemically it is the same element as hydrogen (element number one with one proton) but it has an additional neutron relative to hydrogen; hydrogen has no neutrons, deuterium has one. So deuterium, which is only about 0.015% as abundant in nature as hydrogen, weighs about twice as much as its lighter brother.

The upshot of that mass difference is that the two atoms are readily distinguished in a mass spectrometer. So are molecules containing one isotope versus the other. For example, the mass of ordinary methane, CH_4, is 16 amu, atomic mass units (C = 12; H = 1). If one hydrogen atom is replaced with deuterium, the molecule's mass is 17. If methane has four deuterium atoms, its mass is 20 amu instead of 16.

Our expectation for the hydrogen–deuterium exchange experiment was that if we sent a beam of hydrogen molecules (H_2, also known as dihydrogen, 2 amu) and deuterium molecules (D_2, 4 amu) toward a platinum surface, the molecules would break apart, and some H and D atoms would find each other, form a new chemical bond thereby making H–D (3 amu), and desorb from the surface. But we didn't see any HD molecules.

That result was surprising because platinum was known to catalyze that reaction even at the frigid temperatures of liquid nitrogen. Let me take a moment to explain what that means. Nitrogen, which makes up nearly 80% of the air we breathe, can be liquefied easily. The liquid form of nitrogen looks like water and will remain liquid (usually inside a thermos bottle) as long as the temperature is kept below its boiling point, which is some 320 degrees below zero Fahrenheit (-196 °Celsius or 77° on the Kelvin scale). At that bitterly cold cryogenic temperature, most molecules have very little internal energy and very little molecular "incentive" to react. Yet it was known from other kinds of experiments that hydrogen and deuterium still undergo the H_2-D_2 exchange reaction in the presence of platinum even when the reactants and the metal are chilled to liquid nitrogen temperatures.

To put it simply, even when the molecules have—what would seem to be too little energy to react because they are so cold, the catalytic nature of platinum can still facilitate the reaction because it lowers the energy barrier the molecules must surmount in order to react. So where were our H–D molecules?

One day, Miquel Salmeron, the postdoctoral researcher in my group working on this project, moved the platinum crystal while it was sitting in the path of the molecular beam and suddenly the mass spectrometer detected a large H–D signal. We found that when the molecular beam, which was smaller in diameter than the crystal surface, was aimed at the center of the surface—the way we usually did the experiment, nothing happened. But when the beam fell upon the edges of the crystal, the H-D exchange reaction really took off! What was so special about the crystal edges?

From follow-up experiments, we learned that the method we used to attach the metal crystal to its holder—spot welding—caused a little bit of damage in the welded areas around the edge of the crystal. That process had altered the pristine structure of the crystal lattice and generated numerous defects. Examples of defects include tiny regions in which atoms were "out of place"—like eggs sitting in between the depressions or "dimples" in an egg carton, and lattice positions where atoms were missing. Those kinds of defects (and others) turned out to be catalytically active spots and were responsible for triggering the isotope exchange reaction.

Surface Defects' Role in Chemical Reactivity

We learned a lot of other interesting science in those days by combining molecular beam methods, which probed surface chemical reactions and their rates (or kinetics) with LEED analysis, which revealed the structures of crystal surfaces and the way molecules lined up on them. For example, we learned that one of the platinum faces we studied frequently, the Pt (111) face, was boring chemically. That face of platinum was not much of a catalyst because it did not catalyze any reactions—not even simple ones like breaking the chemical bonds in small molecules.

Now, that observation may sound peculiar. Platinum, after all, is supposed to be a fantastic and active catalyst. Well, for many types of reactions, platinum *is* a fantastic catalyst. But among the many things we learned in those days is that even though every crystal face of platinum is composed entirely of just one chemical element—platinum, the structure of the surface—meaning the exact locations of atoms in the topmost layers—plays a key role in the surface's catalytic properties. Some faces are great catalysts. Others are catalytic duds.

In powdered samples of platinum or other forms made up of tiny particles, thermodynamics of the sample preparation method usually dictates which structures, for example (100), (111), and (110) are adopted by the particle surfaces. If we could analyze a milligram of powdered platinum and tally all of the particles' surface structures and geometries, we would find a large variety present in the sample. We would also find that a few of those structures overwhelmingly dominate the statistics—that is, a few (xyz) labels or Miller indices, as they are known, would show up repeatedly. Some of the ones that predominate are highly active catalytically. And that's one of the reasons that powdered samples of platinum—even though they expose many surface types—tend to be great catalysts.

Another reason, as we would continue to learn, is that defects of various types can impart catalytic activity to faces such as (111), which are inherently inactive. And although our carefully prepared coin-sized single crystals tended to be pristine and defect free, real catalysts—the high-surface-area particulate types used, for example, in the petroleum industry, are full of activity-enhancing surface defects.

We found a silver lining behind the cloud of Pt (111)'s inactivity. The good news is that when we deposited molecules on that crystal surface, they lined up and "waited" for us to study their structures with LEED without reacting. In that regard, the inactivity enabled us to determine the geometries, bond lengths, bond angles, and other parameters of ordered layers of intact molecules on platinum crystals. The bad news, of course, is that these experiments were not teaching us about catalytic reactions.

By the early 1970s we began to get around that problem when we discovered that we could use LEED to study the structure of "stepped" surfaces. To picture a stepped surface, think of a split-level home, where several rooms, for example the kitchen and dining room, are on one level, and the living room is set one or two steps down from the dining room. For crystals that have that kind of structure, surface scientists refer to the horizontal floor area as a terrace, and the vertical portion of the steps (the riser, not the part your foot rests on when you climb stairs) as a step. These steps, which induce a small height offset between terraces, can be as short as one atom. It was a big deal when we figured out how to study them with LEED because surface steps, like other kinds of surface irregularities and defects, are common sites of chemical reactivity.

Not long after learning that stepped surfaces could be studied with LEED, we became interested in a study in which researchers deliberately cut a uranium oxide crystal in a way that exposed a large number of steps and terraces. If you think back to the "stack of egg cartons" analogy of crystal structure, you'll recall that cutting a crystal along certain angles, let's call them "90° angles," exposes the more common faces—the ones with low (xyz) Miller indices. But if crystals are cut at certain oblique angles, it's possible to expose some unusual, highly stepped faces. One of the interesting things about this study, which was conducted by W. P. Ellis and R. L. Schwoebel, is that LEED revealed the intricate structure of the patterned, stepped surface they exposed by cutting it in this uncommon way.

We applied the same approach to platinum and were delighted to find that LEED revealed a complex but nicely ordered surface consisting of steps and terraces. Now we had techniques in hand that would allow us to compare a variety of platinum crystal surfaces. We could prepare largely unreactive surfaces and obtain an accurate picture of the way intact molecules *aligned* upon them. And we could prepare surfaces with steps and related structural features known as kinks and study the ways in which molecules *reacted* upon them. We even had sufficiently fine control over our surface preparation methods to adjust the number or concentration of reaction-inducing defects that appeared on a surface.

In those days, we studied a whole host of chemical reactions. We probed the reactions of hydrogen and deuterium, hydrogen plus oxygen, ammonia plus oxygen, as well as reactions of carbon monoxide, small organic molecules, and other species. Over and over again, we learned about the ways molecules line up on some kinds of surfaces and break apart (dissociate) on other surfaces by severing all kinds of chemical bonds, including the ones in H-H, O=O, C-H, C-C and C=O. It was a successful and rewarding period and one during which my group uncovered a wide range of surface molecular phenomena.

Now is a great time to mention Emery Kozak, a hard-working creative Hungarian man who worked with me as a technician for 15 years beginning in the 1960s. When it came to building high-tech scientific instruments, Emery was a genius. Practically every year that he worked in my laboratory he built a new ultrahigh vacuum system for surface studies. His technical know-how, his original ideas, and his can-do attitude helped fill my lab with high-end research equipment and with a team of students and post docs who acquired some of his instrument-building skills. That knowledge base enabled me to spend my research dollars to great advantage, because it's far less expensive to build these kinds of instruments than to buy them from a vendor.

Emery was a real character. He had been an assistant to military officers in the Hungarian army until the Russians occupied Budapest. Unfortunately, he was captured in one of the Russians' frequent raids and was shipped off to a Russian labor camp where he spent the next three years of his life. Times were tough when he returned to Hungary, and his chances of finding good employment were pretty dismal. So he ended up working as a machinist in a factory there. In the days following the Hungarian Revolution in 1956, he escaped, as I did, and made his way to the United States. As luck would have it, in 1968 when I was looking to hire a technician with excellent machining and design skills, Emery was available, looking for work, and eager to get started. Ours was a marriage made in heaven!

Whenever I came up with an idea about a new a piece of equipment I wanted to build, I would share my thoughts and design plans with Emery. After thinking it over for a while, he would come back and say "You know, we can do the whole thing much better if instead of building the piece of equipment this way, we do it like this and like that." Sure enough, the way he envisioned carrying out the idea that I had presented to him, was better than the way I originally thought it should be done. Thanks to him, we ended up building a whole range of novel UHV systems and various prototype instruments that spurred an era of creative surface chemistry research in my group.

All in all our instruments for LEED, Auger electron spectroscopy, X-ray photo electron spectroscopy, and molecular beam studies gave us a powerful repertoire of analytical instruments to probe surface structures, adsorbed molecules, and catalytic chemical reactions.

We developed this skill set in surface chemistry and catalysis at a fortuitous time with respect to US history. Just as we were building a research group of young scientists who were gaining expertise in those areas, a confluence of factors in US domestic and foreign policies focused considerable attention on research needs for environmental and energy technologies. Surface science was ready and uniquely able to help.

On the energy side, the 1973 oil embargo, which caused the price of transportation fuels to skyrocket and led to hours' long waits at gasoline pumps, drove the US to search for non-petroleum-based fuel alternatives. That event, which was also known as the "oil crisis," was triggered by Arab nations of the Organization of Petroleum Exporting Countries (OPEC) withholding crucial shipments of crude oil to the United States in retaliation for US support of Israel in the Yom Kippur War.

Regarding the environment, air pollution in major US cities such as Los Angeles had increased to intolerable levels. Scientists had determined that automobile emissions in traffic congested areas such as southern California were major culprits in that region's smog and foul air problems. The findings led to the US Congress passing the Clean Air Act and spurred development of the automobile catalytic converter.

Looking back at the 1973 oil crisis with 40 years' hindsight shows that some of the pressing issues of that generation remain issues today. At the core of the discussion then and now is US dependence on foreign oil. As the price climbed back then from $2 per barrel to more than $30 per barrel, the US eagerly sought alternatives to petroleum for making gasoline, diesel, and other transportation fuels. (Nuclear energy was already a success at that time in generating electricity in stationary power plants and in propelling submarines, but was not an option for personal transportation.)

The Energy Research and Development Administration (ERDA), the forerunner of the U.S. Department of Energy, launched a major program to support research aimed at producing synthetic fuels from coal, which is plentiful in the United States. One of the most promising strategies for reaching that goal begins with gasifying coal by treating it with steam and oxygen to produce a gas mixture that is typically referred to as synthesis gas or syngas, and consists mainly of carbon monoxide (CO) and hydrogen. That gas mixture was known as "coal gas" in earlier days and was used widely in the 1800s to fuel gas lamps for street lighting.

Making fuel from syngas means fusing together carbon atoms from CO to form molecules with carbon chains of suitable length. For gasoline, that length is in the range of about five to 10 carbon atoms; for diesel, about 12 to 20 carbon atoms. The most common method for producing these products from CO and hydrogen is by way of a carbon-carbon coupling process known as Fischer-Tropsch synthesis. That process is mediated by metal catalysts—at various times, iron, cobalt and ruthenium have been used. The idea was that by applying the tools and methods of modern surface science to scrutinize the catalyst surfaces that mediate Fischer-Tropsch synthesis, we would be able to elucidate some of the key molecular scale events that control that process. Armed with that information, catalyst manufacturers would ideally be able to make improved catalysts—ones that efficiently convert the starting materials to the desired products selectively (meaning with few side products).

Being well poised as we were to carry out these kinds of studies, my research funding tripled at that time. I embarked on a major program in the general area of energy conversion focusing on syngas catalysis. I also chose to study ammonia synthesis, which even then was one of the older commercial catalytic processes—but one that could benefit from new molecular-scale insights.

The challenge in both cases was that in order to make truly useful discoveries—ones that would be relevant to industrial processes, I needed to emulate typical industrial conditions found in chemical plants. That meant I had to study catalytic surface chemistry at high temperatures and pressures. Initially that seemed quite at odds with the ultrahigh vacuum technology we were pursuing. Even in experiments in which we raised the pressure in the vacuum chamber by a factor of 1000 by admitting a reactive gas, the conditions were very far from industrial catalytic conditions. But it wasn't long before we came up with a solution.

Combined High Pressure-Low Pressure Instruments

To get closer to the conditions that catalysts experience in a chemical reactor—the ones that truly pertain to industrial catalysis, my group designed and built a new type of UHV chamber that featured a small sealable metal compartment within the main vacuum chamber. Good science is often done by adopting the "minimalistic" or "reductionist" approach to experimentation. That is, by holding some (or many) well-understood parameters constant and varying some other parameter, one can confidently tie an observation to the parameter that was varied. Our samples were carefully prepared coin-sized single crystal models of real catalysts—not actual particulate or supported-particle industrial catalysts. By using these kinds of model catalysts, we kept many of the catalyst variables constant and we used our novel hardware to vary other experimental conditions, such as temperature, pressure, and gas composition.

In some studies however, we kept the instrument-controlled parameters constant from experiment to experiment and made changes to the crystals. For example, we compared various crystal faces—including faces with steps and other structural features—to determine which ones actively mediated catalysis and which were inactive. In some cases, we modified the composition of the surfaces with small quantities of additives (or dopants) including various transition metals, metal oxides, and alkali metals, to determine how the rates of catalytic reactions and the distribution of products depend on these variables.

Our high-pressure/low-pressure approach worked like so: first, by applying a battery of surface analysis tools to our carefully prepared and cleaned crystal, we would fully characterize its surface in an initial state in the controlled UHV atmosphere. Then without exposing the crystal to air or other gases, we would seal it in the mini high-pressure compartment. Into that compartment, we had plumbed gas manifold lines, so that after the mini chamber was sealed, we could pressurize it up to a couple of atmospheres (meaning at least two times as high as standard atmospheric pressure or trillions of times higher than UHV pressures) with circulating reactive gases. The sample could also be heated while exposed to those gases, thereby mimicking chemical reactor conditions.

Further, while the sample was in that environment, we could take gas samples from the high pressure chamber via a plumbing line and analyze them with gas chromatography, perhaps the most common method for separating and detecting gas-phase molecules. In that way, we could study the distribution of reaction products that formed and desorbed from our crystal surface and the way that distribution changed over time and with changes in the pressure and composition of the reactive gases. After the reaction portion of the experiment was concluded, we would pump out the high pressure compartment and unseal it. Then, depending on the specifics of the study, we would reanalyze the crystal with the same battery of surface analysis tools searching for signs of chemistry-induced change.

As analytical methods continued to improve—and especially as the theory underlying the physical interpretation of LEED data continued to improve, searching for those signs of chemistry-induced changes to surfaces became more routine. John Pendry of Imperial College in London, who I mentioned earlier, was one of the key forces behind the advancement of LEED theory. I was fortunate to have worked with Michel Van Hove, Pendry's first Ph.D. student, for 16 years. During that period, when he worked in my group as a post doc and later as an LBNL staff scientist, we solved more than 50 surface structures on the atomic scale—meaning that for this large number of distinct specimens, we deduced the relative positions of atoms on surfaces and the angles between the atoms. We also deduced the way organic molecules such as ethylene, benzene, and other important industrial compounds, line up and form molecular bonds on the surfaces of catalytically active metals.

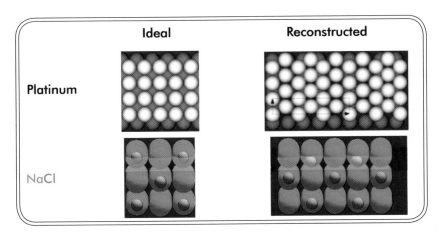

	Ideal	Reconstructed
Platinum		
NaCl		

Surface measurements show that the topmost layers of many
crystals undergo reconstruction—that is, they adopt a geometry
that differs from the geometry of the bulk crystal. These examples
depict ideal and reconstructed platinum and sodium chloride.

Molecular Adsorption

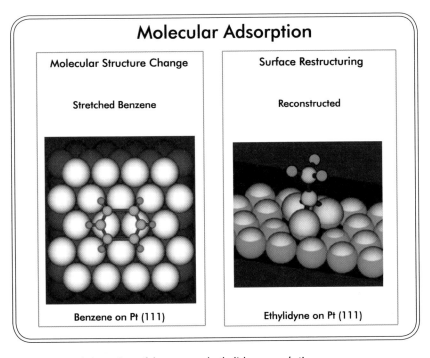

Molecular Structure Change	Surface Restructuring
Stretched Benzene	Reconstructed
Benzene on Pt (111)	Ethylidyne on Pt (111)

Adsorption of bezene and ethylidyne on platinum cause
the metal surface to undergo structural changes.

What emerged from these studies was an understanding of the effects of physical and chemical forces on surface structure. Pristine surfaces (meaning not coated with molecules) typically adopt structures that differ from the bulk because the layer of atoms at the surface sits in a completely different environment than its bulk counterpart. Rather than being buried under thousands of identical layers of atoms, the layers at or near a surface, much like a fish out of water, are highly exposed. Surface energetics dictates the way those atoms "relax" and alter their positions and angles to accommodate that extreme change in environment. And when surfaces are covered with strongly interacting molecules, the molecules drive "adsorbate-induced restructuring," which results in further changes to the structure of the surface relative to pristine surfaces and to bulk layers.

Michel van Hove's calculations and mathematical analysis of our experimental LEED crystallography data showed how pervasive adsorbate-induced restructuring is for many types of adsorbates and surfaces. This phenomenon came to be understood more clearly some years later after the development of an analysis technique known as high-pressure scanning tunneling microscopy.

This combined high-pressure/low-pressure approach to UHV surface science for exploring heterogeneous catalysis became the hallmark of my work for the next 10 years. We built many of these instruments and used them to study a variety of catalytic reactions, such as hydrogenation of CO and CO_2 over iron and rhodium surfaces. Those studies were aimed at investigating syngas chemistry for producing organic fuel molecules. We also studied the surface chemistry of many other hydrocarbon molecules related to fuels. Examples include ring opening (opening the ring of carbon atoms) in cyclopropane; hydrogenation and dehydrogenation of (addition and loss of hydrogen from) cyclohexene; and isomerization (molecular rearrangement) of n-hexane and n-heptane. And we also studied ammonia synthesis on the surfaces of iron and rhenium crystals.

Within a few years, we built up a substantial body of experimental know-how and compiled a large collection of surface science data. Those data, especially from our studies of ammonia synthesis, conversion of n-hexane and n-heptane, and hydrodesulfurization (sulfur removal) from thiophene, served as reference material for many research groups, including ones in industry. And importantly, our findings led manufacturers to reformulate their catalysts to improve performance.

Development of Automobile Catalytic Converters Consulting with General Motors

As I mentioned earlier, the intense air pollution problems in the early 1970s in traffic-congested areas like southern California led to legislation that required automobile manufacturers to curb air-fouling tailpipe emissions. The result was the development of the catalytic converter, which is a mobile catalyst-driven chemical reactor. These reactors are found today on nearly every automobile in the developed world and a large fraction of cars and trucks in the developing world.

Because I specialized in catalytic surface chemistry, which in the early 1970s was not a very widely studied topic, I was asked by General Motors in Warren, Michigan, to serve as a consultant in the development of the catalytic converter. It was a very interesting experience and one that led to many of my students and postdocs being hired by GM.

Catalytic converters have become so effective at cleaning noxious air pollutants from engine emissions, and the technology has been adopted so broadly around the globe, that even people with little connection to science can quickly grasp the motivation for my catalysis research because "catalytic converter" is now a household word. Communicating the importance of science research to non-scientists is important but often difficult. Thanks to the success of catalytic converters, however, it's rather easy to convey to people with little background in science the significance of catalysis in general.

Most automobiles are propelled by capturing the chemical energy released from burning a hydrocarbon fuel such as gasoline or diesel in an internal combustion engine and converting it to mechanical energy. (Battery powered electric vehicles, which may become more popular soon, do not work that way.) The combustion process in a car engine is complex and leads to the creation of a large number of chemical products, some of which are culprits in smog and pollution problems. The variety and distribution of those products depend on a host of variables, including the composition of the fuel and the engine operating conditions.

Burning any kind of hydrocarbon—natural gas, transportation fuels, and candle wax, for example—produces mainly carbon dioxide (CO_2) and water. If we can set aside the discussion about carbon dioxide's role in global warming, which unfortunately has become a highly politicized topic that's often covered inaccurately in the press, tailpipe emissions of CO_2 and water are basically O.K. insofar as they are not toxic. But many other compounds, including for example unburned (unoxidized) and incompletely oxidized gasoline (or diesel), nitrogen oxides (referred to as NO_x), sulfur oxides, and carbon monoxide (CO) are also produced in automobile engines. And those compounds spell trouble. So catalytic converters were designed to deal with them.

Briefly, a catalytic converter consists of a porous ceramic brick that has been coated with nanometer-sized particles of precious metals such as platinum, palladium, and rhodium, and a few other compounds. (Today's versions contain much less platinum than earlier ones did.) The main job of the device is to oxidize CO to CO_2, convert hydrocarbons to CO_2 and water, and convert nitrogen oxides to nitrogen. As a result of those three tasks, these pollution abatement devices came to be known as three-way catalytic converters.

There is much to be said about the success of catalytic converters. But let it suffice to say that they were developed and widely implemented very quickly and that they do their jobs very well and under demanding conditions.

Consider, for example, that they need to *oxidize* hydrocarbons and CO, and (almost) simultaneously *reduce* NO_x. Those reactions are chemical opposites. The conditions that work best for one reaction usually don't work at all for the other reaction. Furthermore, these mobile reactors are repeatedly subjected to enormous temperature swings—say from 0 °F on a winter morning in Minnesota to over 500 °F—in the span of two minutes (or less). In addition, they are constantly jostled, shaken, and rattled by road vibrations and they are expected to work nearly perfectly for 100,000 miles. And in large part, they do!

Credit for developing catalytic converters so quickly in the 1970s needs to be shared among automakers, catalyst manufacturers, mechanical engineers, and others. But without a doubt surface science deserves a substantial portion of the credit. The speedy development would not have been possible without the surface analysis tools that had only recently become available. Those surface science instruments and techniques had only a short while earlier begun providing researchers with detailed information on the structure and composition of catalytic surfaces and on the dependence of their catalytic performance on the chemical composition of air-fuel mixtures and exhaust gases.

This area of research was exciting, fast-moving, and challenging. One especially challenging problem was dealing with the organo-lead compounds, which were added to fuels in the 1970s to reduce "engine knocking" and boost the gasoline octane ratings. The trouble was that the lead compounds "poisoned" or deactivated the catalyst. It took a major effort to get fuel suppliers to eliminate the lead—but eventually unleaded gasoline became, almost exclusively, the only kind of gasoline available.

It's also interesting to recall the way General Motors used its business savvy and international recognition to stabilize the price of platinum, even though demand rose quickly then, by investing in mines in South Africa, which were prime sources of the precious metal. Unfortunately, GM did not remain a leader in the technology or business of catalytic converters. Instead, the company sold the technology to Engelhard and Degussa and even to other automobile companies. The explanation offered was that GM was in the business of manufacturing automobiles, not catalytic converters. Because of my involvement—and that of my students with the company, it was disappointing to witness this turn of events. Nonetheless, it was a fascinating body of research and one that lead to wide scale implementation of technology that directly benefited society and the environment.

Professional Recognition in Surface Science

In the second half of the 1970s, I became increasingly involved with organizational matters of professional scientific organizations and was fortunate to be the recipient of a number of awards. In 1974, for example, I became chairman of the Division of Colloid and Surface Chemistry of the American Chemical Society. In this capacity, I was able to attract and enlist enthusiastic young surface scientists to help increase the popularity and success of this ACS division by organizing timely symposia in important surface science topics. With surface chemistry symposia ranging in focus from biology and medicine to catalysis, electrochemistry, and corrosion, attendance at the bi-annual ACS divisional meetings rose sharply.

Shortly thereafter, honors from various universities and organizations were bestowed upon me in recognition of the work that my research group and I were doing. In 1976, I received the Kokes Award from Johns Hopkins University and was asked to deliver the Phillips Lectures at the University of Pittsburgh. A year later, I was honored by the North American Catalysis Society with the Emmett Award and was invited to Cornell University to deliver the Baker Lectures. And in 1978, I was elected Fellow of the American Academy of Arts and Sciences.

Just one year later I was inducted into the National Academy of Sciences, which is among the highest forms of recognition any scientist can receive from his peers. Indeed, as a 44 year old researcher receiving a professional distinction that was shared, at that time, by less than 200 chemists, this honor was truly outstanding.

In the spring of 1979, Judy and I, together with our children, Nicole and John, came to Washington D.C. to participate in the National Academies induction ceremony. The National Academy of Sciences (NAS) dates back to President Abraham Lincoln, who signed it into existence in 1863. This body was created to "investigate, examine, experiment, and report upon any subject of science or art" when called upon to do so by the U.S. government.

Judy, Gabor, Nicole, and John in 1979

Over the next century, NAS expanded to include distinguished scholars in science, engineering, and medicine who served in advisory roles on the National Research Council and in the National Academy of Engineering and the Institute of Medicine. The National Academies, as the organization is known collectively, continues to this day to advise the Departments of Energy and Defense, the National Science Foundation, the National Institutes of Health, and other government agencies on important scientific, technological, and medical matters.

On our trip to Washington, D.C., we were put up at the Watergate, which at the time, was a very well-known hotel and office complex, because of its connection to the President Nixon era break-in into the Democratic National Committee headquarters. It was a beautiful hotel!

At the main event, my name was called among 60 other scientists who were admitted to the Academy of Sciences as new members that year—and I proudly signed my name in the Academy's honor roll book. It was an enormous honor and a deeply emotional experience that I will never forget.

CHAPTER 7

Catalysis by Model Surfaces

& Construction of Instruments for Molecular Studies of Surfaces at High Pressures (1980-2000)

Surface science was growing up quickly. Measurements that long seemed unreachable and questions that were unanswerable were starting to yield to experiment. Forward march off the beaten path into molecular science!

AS THE STORY I have been telling you so far has unfolded, you can see that for the better part of two decades, my research group built up the know-how to exploit a powerful repertoire of analytical instruments and techniques to probe surfaces, adsorbed molecules, and catalytic chemical reactions. Throughout this period and beyond, the aim was to deepen scientific understanding of surface processes—

especially catalytic surface chemistry—by uncovering the molecular scale events and phenomena that underpin those processes.

The instrument innovations that made this type of research possible have long remained near and dear to my heart. And so I am eager to tell you about the next big thing that advanced instrumental surface analysis.

In a nutshell, our combined low pressure/high pressure techniques—as I described in the previous chapter—were uniquely able to tell us about surface chemistry and surface changes, by carrying out a before-and-after-reaction type of analysis. What we were unable to do, was monitor the surface and the reactions occurring on it, while they were actually occurring—and furthermore, while they were occurring at the temperatures and pressures typically found in industrial chemical reactors. Chances were good that some important processes occurred only under those conditions, and that before-and-after analyses were missing them.

It would be about 10 years until the new techniques that provided solutions to these problems became available. But rather than fast forwarding abruptly to 1990 to tell you about them, let me first share with you some of our research highlights from the 1980s.

If I can borrow a term from today's technology companies, I would say that by making efficient use of our "core competencies" in LEED and combined high pressure-low pressure surface analysis methods, we made significant headway in understanding the nature of numerous solid surfaces. Examples include metals, alkali halides such as sodium chloride and lithium fluoride, iron oxides, and ice. The key findings, as I described previously, are that pristine surfaces undergo reconstruction as a result of the energetics of their "discomfort" in being exposed. Molecules adsorbed on those surfaces cause further reconstruction.

An interesting outcome of this whole body of early surface studies is that the bonding between organic molecules and metal surfaces resembles bonding in organometallic compounds. That observation may seem quite obvious; after all, both cases deal with chemical bonds between carbon and metal atoms.

But it's not that simple. The organometallic compounds in question often consist of a frame or cluster of three or four metal atoms that are bonded to, bridged by, or otherwise interconnected via small organic molecules like CO. All told, those compounds consist of just a small number of atoms.

In the case of an adsorbate, by contrast, a free floating organic molecule lands on and binds to a chunk of metal that in many regards can be thought of as an infinite number of metal atoms. With that picture in mind, it's not obvious what kind of bonds would be expected to attach the molecule to the surface. Recognizing (and proving) that valid analogies could indeed be drawn between metal-surface-bound molecules and organometallic compounds—in terms of the geometry, reactivity, and the general nature of the bonds, for example—deepened understanding of surface chemistry. That way of thinking was one of the new concepts that emerged at that time.

Several other new concepts or ways of thinking about reactions in catalytic surface chemistry were developed in that era. For example, from studies conducted by our group (and other research groups too) evidence emerged that some reactions are quite finicky about the structure of the surface on which they are occur. Other reactions are less fussy.

Ethylene hydrogenation, for example—a reaction that converts ethylene to ethane by adding hydrogen to the molecule, was found to be structure insensitive. That is, the reaction proceeds at the same rate without regard to the structure of the underlying metal surface.

Ammonia synthesis, in contrast, is categorized as structure sensitive because that reaction depends strongly on the structure of the surface of the iron catalyst that drives the process. Some of the rougher, more open iron surfaces facilitate the reaction 100 times more efficiently than other iron crystal surfaces.

Naturally, we wanted to know why that was so. Pieces of the answer came from studies of so called promoters, additives that enhance the rate of reaction. Common promoters for ammonia synthesis include alumina and potassium. It turned out that alumina (aluminum oxide) induces surface restructuring in iron, causing the metal to expose crystal faces that facilitate the highest rates of reaction.

Potassium's affect turned out to be of a different nature. The hallmark chemical property of alkali metals, a group that includes sodium, potassium, and cesium, is their "desire" to shed one electron from their outer electron shell. That property makes those metals—in pure form—explosive on contact with water.

Thus potassium turned out to be an "electronic promoter." Its tendency to donate electrons to its bonding partners weakens the catalyst's hold on ammonia. The weakly held product molecules are thus free to desorb and make room on the catalyst surface for nitrogen molecules to land, bond, and especially—to fall apart (dissociate), which is the key step that controls the rate of ammonia synthesis.

These ideas about structure sensitivity and insensitivity, electronic promoters, and structural promoters, are further examples of molecular-level concepts that advanced understanding of surface phenomena and catalytic chemistry.

We also took up the challenge of trying to understand aspects of catalytic polymerization chemistry. Polymerization of ethylene and propylene to make polyethylene and polypropylene, respectively, is a global scale operation. Those materials are used nowadays in enormous numbers of consumer and commercial products. So studying this topic certainly seemed like a worthwhile use of our resources and one that could potentially lead to an improvement in an industrially-significant area.

That said, I feel compelled to point out that studying this type of chemistry represented a bit of a departure for us from the kinds of gas-molecule—solid-surface systems we had been studying. But venturing into new areas of research is an important—if somewhat daunting—component of building a successful research program. I'll tell you more about that topic later.

Although the names of these products describe their general chemical composition, the names alone—polyethylene and polypropylene—do not tell you very much about the properties of specific products. In short, those labels are names of entire families of polymers in which members differ from one another in terms of molecular size, weight, and structure.

Those kinds of differences at the microscopic scale lead to differences in the products at the macroscopic scale. For example, one formulation of polypropylene may be suitable for use as flexible fibers in fabrics, clothing, and diapers, whereas a higher density and more rigid formulation may be useful for making durable plastic containers or molded into industrial fittings. Polymer manufacturers have long known which procedures lead to which kinds of products, but many of the molecular scale details that guide the polymerization reaction toward one product or another escaped elucidation.

So we undertook a foray into polymerization chemistry with the aim of using surface analysis methods to try to observe molecular scale features that relate to (or dictate) macroscopic materials properties. Through a combination of surface science and catalysis techniques, we watched what happened when we delivered titanium tetrachloride ($TiCl_4$) and aluminum trimethyl, $Al(CH_3)_3$, the polymerization cocatalysts (or more accurately, precursors to the catalyst), to a thin-film catalyst support material made from magnesium chloride and ethanol. As it turns out, we were able to study the catalyst precursors and monitor the composition and oxidation states of the catalytic sites.

Without getting into too many details, let me just mention that one form of polypropylene known as "isotactic" polypropylene is characterized by a structural arrangement in which chemical substituents—molecular appendages—that protrude from the polymer chain or backbone all protrude from the same side of the chain. Isotactic polypropylene tends to exhibit a level of crystallinity that falls between that of low-density polyethylene (known from plastic recycling efforts as LDPE) and high-density polyethylene (HDPE). In atactic polypropylene, which is more flexible and rubbery than the isotactic form, the chemical appendages are distributed randomly.

One of the main results of our studies is that by using surface analysis methods we were able to identify, distinguish, and monitor the catalytic sites that produce isotactic and atactic polypropylene. Some aspects of these polymerization investigations were helpful in designing superactive catalysts endowed with reactive sites that quickly stitch together many millions of carbon atoms.

Overall, the general significance of the body of research we conducted during this decade perhaps can be summed up this way: we advanced an approach to studying catalytic surface chemistry that rests upon using well understood surfaces as model catalysts for mediating industrially relevant reactions. As a result of the observations and discoveries in some of those studies, researchers—often in industry— were able to make chemically rational decisions regarding ways to improve catalytic processes. Those improvements sometimes came in the form of faster acting catalysts or ones that were longer lived, more tolerant of "poisons" and "unfriendly" reaction conditions, or better able to steer a reaction toward a desired product while producing little or no waste products.

The so-called rational-design approach is inherently more efficient than the classic trial-and-error or "let's try it and see what happens" approach that long characterized catalysis research. That's not to downplay the value of an experienced scientist's chemical intuition. That type of talent and scientific wisdom is invaluable. But chemical intuition coupled with molecular scale insights into reaction phenomena is more powerful than chemical intuition alone.

One reason I mention that some of the results of our studies were put to use by scientists in industry is because of the consulting work I did for companies such as General Motors, Union Carbide, Dow, Imperial Chemical Industries (the British chemical company better known as ICI) and Montecatini, a large Italian chemical company. The automotive emissions catalysis work we did with GM, the hydrocarbon catalysis projects with Union Carbide, and the ammonia synthesis work for ICI, for example, were interesting and exciting projects. I particularly enjoyed working with Jule Rabo at Union Carbide, with Ken Waugh at ICI, and Paolo Galli at Montecatini.

As an academic scientist, however, I was always aware of the limiting effects of the confidentiality agreements under which consulting research was carried out. In short, research success in academic science is measured in part, in terms of the quantity and prestige of a researcher's scientific publications. But consulting work typically is kept confidential, so many of our results from those studies were never published in the open scientific literature. Yet it is clear in many ways, including the duration of these consulting relationships and the frequency with which my students were subsequently hired by these companies, that a number of our consulting results and openly published results led to improvements in industrial catalytic processes.

Despite our overall success in making headway into elucidating molecular scale surface reaction phenomena, we remained beset by a major characterization flaw. As I pointed out at the beginning of this chapter, we were generally unable to apply our analysis methods to surfaces while they were mediating reactions—and more importantly—while they were doing so under typical industrial catalytic reaction conditions. Those conditions include catalyst surfaces exposed to high temperatures and high pressures of reactive gases or in contact with reactive liquids.

The key reason for that analytical shortcoming is the general incompatibility of those kinds of conditions with ultrahigh vacuum, which is the condition under which most of our analytical tools worked best. It was clear to us that some chemistry processes can only occur readily at the interface of a catalyst surface with a high concentration of reactant molecules. I say "readily" because there is some probability, albeit infinitesimal, that if there is just a handful of molecules sitting on a catalytic surface, one of them may undergo reaction while it is being observed.

Increasing the concentration (the gas pressure or "number" density) of reactant molecules by a factor of 1 trillion, for example, can raise the probability of observing a reaction by a correspondingly large factor. Until this point, however, we did not have the means to analyze surfaces under such high pressures. By examining catalytic surfaces—as we did at that time—only before and after reaction, we were, no doubt, missing out on important observations.

So my group and I embarked on a quest to develop surface analysis techniques that could be used at high pressures. Together with our colleagues, we developed two such techniques in the 1990s: high-pressure versions of scanning tunneling microscopy and sum-frequency-generation vibrational spectroscopy. We also made use of a third high pressure method—a variant of X-ray photoelectron spectroscopy—that was developed at Berkeley during that period.

Scanning tunneling microscopy (STM) is one member—the original—in a family of scanning probe microscopy methods. As its name indicates, this collection of methods yield micrographs—routinely at a magnification of 1 million or more—by scanning (or rastering) an incredibly sharp probe tip across a surface. STM was invented in the early 1980s by Gerd Binnig and Heinrich Rohrer of IBM Zürich and the pair were honored for this game-changing analysis tool with the 1986 Nobel Prize in Physics.

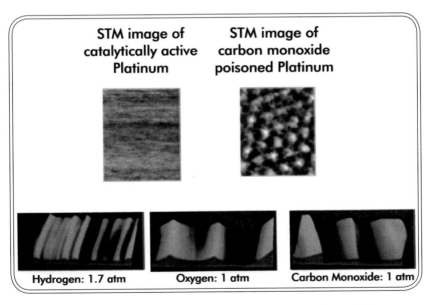

The presence of hydrogen, oxygen, and carbon monoxide on platinum cause the metal's surface to undergo reconstruction, as revealed by these high-pressure scanning tunneling microscopy images.

The basis of detection in STM is completely unlike that in conventional light microscopy and transmission electron microscopy. In those two methods, magnified images are formed through the actions of lenses—like the ones in reading glasses—that collect and focus beams of light (or electrons) that are deflected from (or transmitted through) a specimen.

In STM, in contrast, a clever electronic feedback system brings the incredibly sharp tip to within a few atomic lengths of a surface. As the tip scans the surface, an odd quantum mechanical phenomenon (meaning one that does not truly exist in our macroscopic world) causes electrons to tunnel from the tip to the sample or the other way around depending on how a voltage (or bias) is applied to the tip and sample. The feedback system can be set to constantly adjust the height of the tip above the surface in a way that maintains a constant tunneling current. In so doing, the tip follows the atomic-scale hills, valleys, troughs, and other surface contours thereby revealing the surface topography at the angstrom level.

STM, which was originally used in vacuum, is especially adept at revealing the locations and structures of subtle features such as atomic steps, kinks, and other kinds of irregularities or defects, on an otherwise infinite and well-ordered crystalline landscape. The technique can also pinpoint adsorbed atoms and molecules provided that they hold still long enough to be imaged.

We built a specialized scanning tunneling microscope (I'll skip the instrument details) that could be used to image surfaces under pressures ranging from ultrahigh vacuum to a few atmospheres and at temperatures of up to around 200 °C. Under those conditions, which were far closer to typical chemical reactor conditions than we had come previously, we closely monitored the way surfaces reconstruct upon exposure to reactive gases such as hydrogen, oxygen, and CO, and could discern specific differences in the modes of reconstruction. The method showed us features such as missing rows of atoms and various types of microfacets that formed only under high pressures of select gases.

We also learned that adsorbates are not glued to the surface. In fact, they tend to be quite mobile and diffuse across the surface too quickly relative to the tip scanning speed to be imaged. We also found from these high-pressure STM studies that we could add a second type of adsorbate that tends to pin the first type in place. The studies show, that as a result of the reduced freedom to diffuse, the adsorbed molecules form ordered structures that stay put long enough to be imaged. Not only that, the molecules stop undergoing reaction because they no longer have the mobility to zip around the surface and find catalytically active hot spots needed to perpetuate the reaction.

STM and the related scanning probe methods, such as atomic force microscopy, have remained analytical methods of choice for many researchers since they were commercialized in the 1980s. As a proud surface scientist, I'm delighted that these scanning probe microscopes, all of which are surface analysis tools, have become incredibly popular and useful research instruments.

Just to mention, some of these instruments are routinely used by scientists who would be unlikely to classify themselves as surface scientists. Just the same, the analyses they conduct with these techniques always focus on surfaces. Polymer scientists, for example, routinely evaluate physical properties of novel types of thin films of organic polymers via atomic force microscopy (AFM). And other researchers, working in unrelated areas such as engineering alloys and biominerals, also use AFM to measure hardness, strength, and other properties of surface layers relevant to their fields.

143

To this day, some of my former students and post docs continue to push the methods' capabilities to advance what I'll call classic surface chemical-reaction analysis. Their aim is to continue to advance these techniques to generate atomic resolution images and other atomic-level information at still higher and industrially-relevant pressures and temperatures while the reactions are occurring on reactive surfaces.

The other high-pressure analysis technique we worked on at that time is known as sum-frequency generation (SFG), which is a surface sensitive vibrational spectroscopy method. SFG belongs to a family of laser-spectroscopy techniques that take advantage of unusual processes in a field known as nonlinear optics to interrogate molecules. My colleague Ron (Yuen-Ron) Shen, a long time professor in Berkeley's physics department, was a pioneer in this area of optics and laser spectroscopy and we worked together to develop the methodology needed to apply SFG to surface chemistry.

Briefly, in the SFG process, two beams of laser light, each with a distinct frequency, interact at an interface in a way that leads to a seemingly bizarre result referred to as up-conversion. Up-conversion is a radiation mixing process that generates a beam of light with a frequency equal to the sum of the frequencies of the two input beams. This process of mixing frequencies, which are related to the light beams' wavelengths and colors, is somewhat similar to mixing two paint samples on a painters palette and generating a third color. Of course, human eyes only see colors in the visible portion of the electromagnetic spectrum. But in principle, this analogy holds throughout the entire spectrum.

The way we carried out the SFG experiment was like so: at the same time that we irradiated a molecule-covered surface with visible laser radiation, we irradiated the same sample region with infrared laser light as we scanned the IR laser's frequency. By irradiating with one fixed-frequency beam and one scanned-frequency beam—and by tuning the IR laser to select frequency ranges—we were able to measure the adsorbed molecules' vibrational spectra (which can be a treasure trove of molecularly specific information) thanks to a quantum peculiarity of this non-linear optical process.

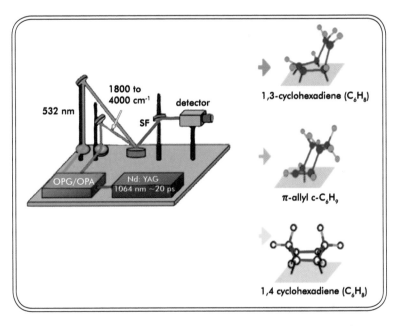

Sum frequency generation (SFG) vibrational spectroscopy can be used to identify and analyze adsorbed molecules by measuring their vibrational frequencies. In an SFG apparatus (left), the frequency of the visible laser beam (green) is fixed, and the IR-beam (red) frequency is varied. The technique is surface sensitive—that is, it only probes molecules adsorbed on the surface. The technique revealed three short-lived reaction intermediates in cyclohexene hydrogenation/dehydrogenation on platinum (right).

It turns out that due to quantum "selection rules," those are laws of nature in the quantum world that dictate what's possible and what is not, the SFG process cannot occur in an isotropic medium—i.e. a highly symmetric environment such as the bulk of a solid, liquid, or gas. So if we see manage to detect a sum-frequency signal (which was not so simple in the early days of this experiment), it had to have originated from the surface. The reason for the surface specificity is that the surface is inherently asymmetric; the layers below a surface (the interior of the solid) are different than the layers above, where vacuum or adsorbed molecules reside.

Armed with this innovation, we now had the power of vibrational spectroscopy, which had long been used for qualitative and quantitative analysis of an enormous number of organic and inorganic molecules, in a surface-only format. By "surface only," I mean the SFG surface-species data would not be buried among bulk-species data. That selectivity or surface specificity is key. But even more than just being a surface-sensitive version of a traditional and powerful chemical spectroscopy tool, SFG opened new possibilities in surface research as a result of its most basic nature. SFG spectroscopy is a light in–light out technique.

Perhaps you remember a couple of chapters back when I began describing vacuum science and the necessity of conducting many types of surface studies under ultrahigh vacuum conditions. You may recall that I made an analogy to the futility of trying to play baseball under water. I told you that at high pressures, many surface science techniques simply cannot work because the electrons, ions, atoms, and molecules that need to travel—often several centimeters—to a sample and/or to a detector get deflected along the way and cannot reach the sample or the detector. But SFG spectroscopy does not depend on those particles. It depends on light. And light is not deflected (much) under those conditions.

We soon began applying SFG spectroscopy to interrogate the structures, compositions, and orientations of monolayers of molecules on surfaces. As my group's experience with this method increased, we learned how to probe molecular layers at a variety of types of interfaces including gas-solid, liquid-solid, and gas-liquid. The technique was applied to deduce what was happening in reactive systems under high temperature and high pressure and to learn how those systems responded to changes in those parameters. Unlike vacuum-solid interfaces, which we had studied for many years—and which were excellent model systems, these other interfaces are "real world" interfaces where most reactions (on Earth) take place.

We found SFG spectroscopy to be a highly versatile tool that could be used effectively to uncover molecular-level information in a wide range of chemical systems. Naturally, we applied the method to study catalytic reactions of hydrocarbons because of our long standing interest in that area. Within that category, we studied hydrogenations, oxidations, isomerizations, and other types of rearrangements in many classes of molecules. Some examples include alkanes, alkenes, and alkynes (compounds with single, double, and triple C-C bonds, respectively); aromatic compounds such as benzene; and a variety of linear and cyclic compounds.

Through these SFG investigations of catalytic reactions under conditions that mimic the environment inside industrial chemical reactors, we found evidence of short-lived reaction intermediates. Intermediates are molecular species that briefly come into existence as reactant molecules undergo transformation to products. These species, which tend to be invisible to many analytical methods, often exhibit unusual bonding and chemical structures—but only fleetingly. Detecting them and understanding their nature is often the key to improving a chemical process.

In 1987 while I was briefly visiting at the ETH in Zurich, Switzerland, Balzer Science Publishers invited John Meurig Thomas of the Royal Institution in London and I to serve as coeditors-in-chief of *Catalysis Letters,* a new journal that they were launching at that time. We accepted the offer and the journal's first issue was published in 1988.

25 years later, *Catalysis Letters* and its sister journal, *Topics in Catalysis,* are still successful and running strong and I still thoroughly enjoy serving as editor. The two journals have helped foster rapid global expansion in the field of catalysis science not just in the U.S. and Europe, but in Asia and South America too. Recently, Norbert Kruse, a professor of catalysis and surface science at Free University of Brussels, stepped in to serve as coeditor-in-chief following Prof. Thomas' retirement. In 2013, Hajo Freund, director of the Fritz Haber Institute of the Max Planck Society, in Berlin, took over the reins from Prof. Kruse as coeditor-in-chief.

Surface Science Applications in Electrochemistry, Polymers, Tribology, and Biointerfaces

Catalytic reactions related to hydrocarbons, fuels, and petrochemistry was but one area that benefited from SFG spectroscopy. Several other areas tied to surface science benefited too. Electrochemistry, for example, which is the science at the heart of all batteries and fuel cells, has attracted the world's top scientific minds for more than two centuries. Due to rapidly expanding interest in hybrid and electric vehicles and ever more powerful portable electronics, electrochemistry is as hot today as ever.

The field's lengthy history is rich with examples of researchers who designed systems that generate electrical power from chemical reactions. The unifying thread in these system designs is that the reactants are kept separate from one another, and the transfer of electrons from one reactant to the other—that's the hallmark of these types of reactions—occurs through an external circuit.

The transfer or flow of electrons is electrical current that flows from one electrode in contact with one reactant, through some kind of device—a light bulb, iPod, or cell phone, for example—to the other electrode which sits in contact with the other reactant. Many such electrochemical systems are reversible. The "forward" reaction converts reactants to products and generates electricity. In the "reverse" reaction, a cell phone user provides electricity to charge the phone's battery, which transforms products back to reactants so the electrochemical cycle can start again.

Researchers have developed many techniques over the years for studying electrochemical reactions. But with the advent of SFG spectroscopy, and in particular, the method's ability to probe the interface between a metal electrode and a liquid-phase reactant, we had an opportunity to test the waters and see if we could apply our new surface analysis method to an area outside of our bailiwick. Phrased a little differently, SFG spectroscopy provided us with an opportunity to examine an area of surface chemistry that had traditionally been considered off the beaten path for surface science. But surface science is exactly the right category for these chemical processes because they are electrochemical reactions occurring on electrode surfaces.

So together with Ron Shen's group and Phil Ross, a Lawrence Berkeley electrochemist, we set up a test cell in which we could probe (by mixing the requisite laser beams) platinum electrodes while they were in contact with organic liquids. In one such experiment, we used SFG spectroscopy to watch a small organic molecule, acetonitrile (CH_3CN), flip-flop on the electrode surface in response to an applied voltage.

We found that at low voltage the molecule tends to stand with its methyl (CH_3) group on the electrode surface and CN group "up in the air." At higher potential, the molecule flips around. The beauty of the technique is that we could use it to watch these molecular acrobatics unfold reversibly as we ramped up and then reduced the applied voltage.

Another non-traditional surface science topic we tackled with SFG spectroscopy is the surfaces of polymers. Earlier I told you how we studied the surfaces of polymerization catalysts to understand the mechanism through which they can impart structural and other properties to synthetic polymers. At this point in the story, we studied the surfaces of the polymer products themselves.

For example, we applied SFG spectroscopy to learn about the surface structures and compositions of commercial polymers such as polyethylene, polypropylene, and polyurethane. We studied the differences in the way these polymer surfaces—meaning the molecular chains at the surface of polymer samples—reorient themselves in response to the presence of air or liquid water. We also examined the surfaces of polymer blends and found that quite often these seemingly homogeneous blends tend to segregate at the surface. That kind of molecular segregation is analogous to strands of cooked whole-wheat (dark) spaghetti spontaneously wending their way to the top of a pot filled with a mixture of whole wheat (dark) and regular (light). Interestingly, we found that polymer segregation in blends was the rule rather than the exception.

Another branch of surface science we concentrated on is known as tribology. Tribology is the science of friction, wear, and lubrication—and even if the subject name is new to you, the effects are not. In nearly every case in which macroscopic materials (big enough to see with the naked eye) slide past one another, the sliding and rubbing action stimulate friction, heat the materials, and cause wear. If you think about it for just a moment, it will be clear that all of this activity must happen right at the surfaces that are in contact with one another—the so-called sliding contact. These everyday processes slowly take their toll on moving parts everywhere; in engines, in bridges and roadways, in the components of door hinges and locks, and on the soles of every pair of shoes ever worn for walking or running.

Have you ever wondered why ice is slippery? That question, which has puzzled scientists since at least the mid-1800s, falls squarely in the domain of tribology because slipperiness is dictated by friction and lubrication. Over the years, researchers have proposed a variety of explanations for ice's slipperiness—for example, melting induced by the pressure of an ice skater's blade or as a result of friction between the blade and ice. But only with the advent of modern surface science tools, including LEED and SFG spectroscopy, have scientists been able to confirm or disprove those proposed explanations.

In a nutshell, surface analysis shows that even at temperatures far below water's conventional (or bulk) melting/freezing point, a molecularly-thin layer of liquid-like water remains on the surface of solid ice. That may seem strange: how can water remain liquid below its freezing point? The answer is that the water layer is an ultrathin quasi-liquid. It forms on the ice surface so as to minimize the surface free energy. In simpler terms, nature "prefers," energetically speaking, to retain this thin quasi-liquid layer on the surface of ice even when it's cold. This phenomenon is referred to as surface premelting—and it isn't unique to water. Among other things, we learned from our SFG studies that this liquid layer thickens with rising temperatures and with the presence of air bubbles and impurities in water, all of which directly impacts the experience of hockey players, figure skaters and everyone else who walks or drives on ice.

In addition to performing what I'll call classic (yet modern) tribology studies using an atomic force microscope (AFM), we also used this type of microscopy, sometimes in conjunction with SFG spectroscopy, to study tribological properties of polymers. By "classic" studies, I am referring to investigations of conventional engineering materials such as metals, oxides, and various types of carbon including graphite, diamond, and other hard carbon materials.

In these experiments, an ultrafine AFM tip, in some cases made of silicon nitride, is lowered with incredible precision until it just makes contact with the surface. Then it is slowly dragged across the surface while a feedback system measures the forces that deflect the tip during dragging. The setup can be used to measure a variety of materials properties such as friction coefficients, stiffness, elasticity, hardness, and the minimum load necessary to break chemical bonds between surface atoms.

We performed related AFM tribology experiments on polymers. And by coordinating those measurements with SFG analysis of the same surface regions, we were able to relate the surface structure of polymers such as low-density polyethylene to that material's mechanical properties. In one example, we combined the two methods to compare pure low-density polyethylene to a commercial sample of that polymer. The SFG spectra of the two samples looked markedly different, with the commercial sample uniquely exhibiting a large concentration of surface methoxy ($-OCH_3$)-containing additives. Judging from results of AFM analysis, the additives serve, at least in part, to reduce the pristine polymer's stiffness, adhesion, and friction.

In much the same way that solid–vacuum and solid–gas interfaces are tied conceptually to more complex interfaces, such as those of synthetic polymers, all of these interfaces are tied by the same concepts to biological interfaces—think of it as the next frontier. As a result of ongoing advancements in surface analytics, surface science opened its grasp widely during a period of just a couple of decades, growing from high vacuum surface science to high pressure surface science, to liquid-solid surface science, and to the surface science of man-made polymers. This trend drove the field to help uncover the molecular nature of systems found in everyday life. Now the field was poised to try to answer questions about the nature of complex interfaces inside living systems.

Surgically implanted devices have broadly improved quality of life for many people and have significantly increased human life expectancy. While it's difficult to point to implantable devices or any other specific medical advance as the cause for increased longevity, longevity has indeed increased. According to U.S. Census Bureau data, life expectancy for American males increased from roughly 67 to 76 years from 1970 to 2010. In the same period, the numbers for U.S. women rose from 74 to 81. Surely, implantable devices play a role in this trend.

Stents help open clogged arteries, fresh heart valves replace weak ones, implantable defibrillators help hearts beat regularly, and engineered knees and hips take the place of those body parts in cases where they are worn out. For patients to benefit from these surgical procedures, however, their bodies must accept the polymeric or polymer-metal-based replacement parts. That is, these materials must be biocompatible, which means their surfaces (and here is the connection to surface science) must be able to support continued cell growth and physiological integration. Even non-surgically-implanted products, such as contact lenses, also must "agree" with the body.

In many areas of scientific research, simplifying the system under examination and reducing the number of variable parameters is the key to sorting out what's going on. It certainly is the case in biointerfaces. To make headway into the complex field of biointerfaces or biocompatibility, it is helpful to regard the internal surfaces of human bodies simply as an interface between biological polymers (in membranes and tissues) and water (the solvent of biological fluids).

With that idea in mind, we conducted a basic study of biocompatible polymer blends that were being evaluated for use in intravenous catheter applications to determine the nature of the materials' surfaces. The polymers were polyurethane-based materials capped with poly(dimethylsiloxane) (PDMS) and fluoroalkyl groups.

Through a combination of SFG analysis and XPS, we found that the composition of the polymers' surfaces was noticeably different than their bulk. As with other polymer blends we studied, these biocompatible blends responded to the interfacial environment by enriching the polymer surface in a single component. We found that the surface chemistry of such blends (which in large part controls biocompatibility) could easily be dominated by a minor component. That's an example of the kind of information can be helpful for screening and selecting candidate materials for various medical applications.

All of these fields that at one time would have been considered on the fringe of surface science continue to march forward today. Complex biological interfaces, where proteins, peptides, and amino acids play subtle but crucial roles, are slowly giving up their molecular secrets to SFG and other probe methods. These techniques have revealed the precession of lipids across lipid bilayers and have nailed down that absolute molecular orientation of large biomolecules, thereby helping to elucidate the mechanisms by which these species mediate key biological functions such as cell growth.

It is simply astounding to consider how broadly the reach of surface science has grown in a relatively short time. Complex systems that would have seemed untouchable in the early 1990s were quickly becoming accessible to molecular probes by the late 90's. And the trend has continued. A number of my students and post docs, as well as gifted scientists who did not work with me, have continued to develop scanning probe and other microscopy tools, laser-based methods, and various spectroscopy techniques to look ever more closely at molecules that reside at interfaces and to deduce their nature and function.

Challenges of Changing Research Directions

The 1980s and 90's was a period of rapid growth and maturation in surface science. Cleverly designed analytical instruments made it possible to observe a host of molecular scale phenomena in several technologically important areas. In the preceding section, I gave you just a taste of the progress made in understanding the basics of petrochemical and polymerization catalysis, tribology, electrochemistry, polymer surface properties, and biological interfaces. It was exciting to witness this progress and to be able to lead a team of such productive scientists through this period. But it was also a bit nerve racking. It seemed that every time our research hit a stretch of smooth sailing, we needed to consider rocking the boat, taking risks, and shifting directions.

Ever since my early days as a junior researcher in the 1960's, my dream was to understand the chemistry of surfaces at the molecular level. Early on I chose surface catalysis on metals as my main subject area and focused on single crystals because I could prepare them reproducibly with known structure and composition in high vacuum. My tools were LEED and molecular beam scattering and I used them to study simple molecules.

155

We soon moved in two directions simultaneously; toward detailed studies of the nature of adsorbed molecules, and toward investigation of surface processes via combined high-pressure low-pressure surface analysis systems. Soon thereafter I had to abandon our molecular beam surface scattering work in order to fund our studies of catalysis at high pressure. As we embarked on advancing STM and SFG instrumentation to probe surfaces under reaction-like conditions, we needed to abandon our LEED work, and several of our combo high-pressure–low-pressure systems were soon disabled.

Every time we identified a critical shortcoming in our methodology for deducing surface molecular phenomena or uncovered a new opportunity for advancing surface science, we excitedly set about finding a new way to answer unanswered questions. After we began making headway into analyses of solid–liquid interfaces, for example, we wondered whether to continue in that direction and explore topics such as biointerfaces and electrochemistry—and move away from the solid-gas interfaces of heterogeneous catalysis. For a while we tried to work in both areas but it proved difficult to remain at the leading edge of research in both subjects.

More recently we began considering whether to exploit the growing body of nanomaterials synthesis to prepare new types of model catalysts; custom nanoparticles. As I'll describe to you in the coming pages, the explosion of nanoscience information provides a perfect opportunity to prepare a great variety of materials as individual particles with diameters in 1-10-nm range. It turns out that size range is incredibly important in nature and industry.

Would it be prudent to invest my group's time and resources mastering these preparation methods in order to try to make new discoveries in surface catalysis? Or would we better off focusing on the high pressure tools that had been serving us so well. Once again it was decision making time. I'll cut the suspense and tell you that nanocatalysis is what we focused on next.

The journey through changing research topics forces us to make critical decisions about where to concentrate our effort and what to leave behind. The entire process keeps our research program fresh and invigorated but certainly causes a lot of anxiety. I'm not complaining. It's merely a bit of work stress that comes with the territory!

Awards and accolades

1980 to 2000 was a highly productive time for my group. I trained a large number of graduate students and postdoctoral fellows who went on to successful professional careers in academia and industry. I wrote another textbook, "Introduction to Surface Chemistry and Catalysis," which was published in 1994. This text is aimed at upper level students in chemistry, materials science, and engineering, and focuses on atomic and molecular properties of surfaces. The book draws heavily on discoveries made during the preceding three decades.

In 1994 I was invited to the University of Cambridge, in England, to deliver the first Linnett Lectures. Judy and I then moved on to Oxford University where I was invited to give the Hinshelwood Lectures. We stayed there for more than a month and had a wonderful time. Sadly, my mother passed away while we were in England, so we returned to Berkeley promptly for her funeral.

My mother was a hardworking woman who had tremendous inner strength. For many years, she worked side-by-side with my father in their retail business. She was completely dedicated to her family and did her very best to provide for us under terrible circumstances, such as the period during the Second World War when my father was forced to serve in a Hungarian army work detail. Through her actions, she taught me life lessons, especially on the importance of family commitment.

During this period, I received several professional awards that reflect the importance of molecular surface science and the field's recognition by my peers. From the American Chemical Society, I received the Colloid and Surface Chemistry Award (1991), the Peter Debye Award in Physical Chemistry (1989), the Adamson Award in Surface Chemistry (1994), the Creative Research in Catalysis Award (2000), and the Linus Pauling Award (2000). The Materials Research Society honored me with the Von Hippel Award (1997).

I received honorary doctorates from the Technical University of Budapest (1989) where I was an undergraduate before escaping to the West after the failed Hungarian Revolution of 1956. Later, in 1992, I was made a foreign member of the Hungarian Academy of Sciences. And I was recognized with honorary degrees from the Université Pierre et Marie Curie, Paris (1990), Université Libre de Bruxelles (1992), University of Ferrara (2000), and the Royal Institute of Technology in Stockholm (2000).

In 1998, I received the Wolf Prize in Chemistry, sharing the honor with my contemporary and longtime friend Gerhard Ertl of Technical University of Berlin and the Fritz Haber Institute (FHI). Ertl had spent a sabbatical leave in my laboratory in 1979 and I spent time at FHI as a Humboldt Senior Fellow in his laboratory in 1989 just before the Berlin Wall came down.

The Wolf Prize was presented by Israel's president at a lovely ceremony at the Knesset (the Parliament Building) in Jerusalem. I was certainly grateful to have received the earlier prizes. But being acknowledged by this prestigious non-American award made a powerful statement about the importance of my scientific work and its recognition internationally. I was truly honored.

The trip to Israel was wonderful. In Jerusalem, my family met Micha Ascher, who had been a postdoctoral fellow in my group in the 1970s and had since become a professor at Hebrew University. He was my first Israeli postdoc. We also met with Raphael D. Levine, a Hebrew U. chemistry professor and theoretician and longtime friend. After the award ceremony, we enjoyed a fascinating tour of Jerusalem and had a chance to visit many beautiful and historically interesting sites in Haifa, Jaffa, Tel Aviv, Bethlehem, and other landmark locations in Israel. All in all, the visit to Israel was unforgettable, touching, and emotional. It left me with warm memories and a deep appreciation of the Jewish homeland's long and rich history.

Chapter 8

Nanocatalysis

(2000—2010 and Beyond)

Nature accords special favor to things in the nanometer size range. That's especially so in catalysis. The more we learn about the traditionally distinct disciplines within that field, the more similar they seem. Like the nanometer sized bits of metals on a support material that comprise heterogeneous catalysts, so too the chemically active hotspots of molecular (homogeneous) catalysts and biocatalysts (enzymes) are also nanosized.

IT ISN'T EASY to pinpoint on a timeline precisely when it happened, but somewhere around the beginning of the 21st century, there was a colossal explosion in the world of nanometer-scale science. Seemingly overnight, researchers everywhere began demonstrating incredible feats of nanoscale dexterity—and the news began spreading very quickly.

The explosion was not confined to the corridors of academies and scientific institutions—though it certainly was felt there most strongly. In laboratories all along those hallways, scientists devised clever techniques for positioning—manipulating—tiny numbers of atoms and molecules, sometimes just one at a time, placing them where they wanted them, forming patterns, and often, mostly for effect, writing famous quotes with nanoscale letters that then they read back through micrographic imaging.

Other researchers came up with methods for synthesizing an assortment of nanoscale dots of prescribed size and composition, solid nanowires of various materials, and hollow nanotubes. This extreme adroitness led to a host of never-before-seen nanometer-sized particles of wide ranging shape and complex composition. At the same time, the microelectronics industry continued its long trend of reducing the size of integrated circuit features via nanoscale lithography methods. Science had begun to get a handle on complex molecular scale patterning.

Outside of the laboratory, news reports, T.V. specials, and a host of other media outlets were buzzing with stories about these dexterous developments. Curious people with little or no connection to science were fascinated by this area of physical (and soon biological) science, and with little warning, "nanotechnology" and the prefix "nano" became everyday words.

Meanwhile, back in the world of heterogeneous (or surface) catalysis, many of my colleagues were quick to point out that nanoscale entities, in and of themselves, were nothing new. Heterogeneous catalysis had long been firmly anchored in nanoscale materials science. For decades, the microscopic species that drove industrial catalytic chemistry were metal particles in the low nanometer size range. Nanosized metal clusters, typically supported on—or in the pores of—high-area support materials such as alumina and silica, were—to put it bluntly—"old hat" in catalysis. Nanoscale patterning, however, was *not* old hat—and so now we had an opportunity to take advantage of these advances to push forward in our understanding of elementary surface chemistry.

Guided as my group had been for decades by the reductionist approach to experimentation—the strategy of reducing the number of variables and irrelevant parameters to emphasize the important ones—we turned to electron beam lithography in this period to fabricate arrays of catalytic nanoparticles.

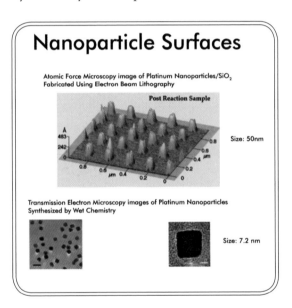

The idea was to make new types of model catalysts—not the (relatively) large and simple low surface area single-crystals we had studied for years, but rather, row upon row of neatly aligned, uniformly sized, and uniformly spaced metal nanodots of identical chemical composition. In principle, such models would closely emulate real-world catalysts (I'll explain in a moment) and eliminate catalyst-to-catalyst variations in surface geometry, predominating crystal face, electronic structure, and variations in other factors that could alter chemical reactivity and selectivity.

By fully characterizing these new model catalysts and using them to study test reactions, we aimed to learn more about the fundamentals of catalytic chemistry and more about the key factors relevant to industrial processes than had been possible in studies based on single crystals. That aim was rooted in the strong similarity between these new models and real-world catalyst, which often are oxide-supported metal particles with diameters in the 1–10-nm size range.

We chose electron beam lithography and the closely related method of photolithography because they were well established techniques used widely in the microelectronics industry for making large numbers of identical finely detailed patterns. (They still are.) I'll spare you the details. The key points are as follows: a computer controlled beam of electrons (or photons in the case of photolithography) writes a fine pattern into a film of a radiation sensitive material known as a resist. The pattern forms as a result of exposure to the radiation. By following the exposure step with a chemical treatment, the exposed portion of the resist can be washed away leaving behind a stencil (a cutout template).

The template can be used repeatedly and in more than one way. For example, in some studies we evaporated a platinum film uniformly across a template that had been riddled with an ordered set of holes—and then removed the template. In that way, the underlying surface, for example alumina, was easily decorated with billions (or more) of identical platinum particles precisely—and only—where the holes exposed the surface.

Our early studies in this area verified that these arrays of metal nanodots were, first and foremost, viable model catalysts. Microscopy analysis proved that they had turned out to be highly uniform, just as we had planned. Spectroscopy analysis and catalysis tests based, for example, on ethylene hydrogenation, showed that in ways that do not depend on particle size or surface area, these arrays behaved very similar to our single crystal samples. That outcome was to be expected, of course, but had to be proved. But although these nanoparticles were on the order of just 15 nm tall, 30 nm in diameter, and separated from their neighbors by about 100 nm, they were still quite a bit larger than the 1–10-nm size range we were aiming for.

So we pushed on in nanocatalysis and experimented with photolithogpraphy and other fabrication methods. But those methods also left us somewhat short of our low-nanometer size goal. So we turned instead to colloidal chemistry methods for preparing metal nanoparticles.

Colloidal chemistry is an old field. Without worrying about detailed definitions, let me say simply that "colloid" typically refers to a suspension of microscopic particles (solids) in a liquid. Some scientists call the particles themselves "colloids," which can be a little confusing. Naturally occurring colloids are common. Suspended mineral particles and organic or biological matter in lake or river water are examples. Milk and blood can be considered colloidal systems in so far as these complex fluids contain organic and biological particles suspended in a liquid medium. The cleaning action of many laundry detergents is based on the tendency of complex colloidal ions known as micelles to wrap themselves around dirt particles and lift them from a stained fabric into the wash water. And colloidal particles also get the credit for the beautiful coloring of opals.

A large number of studies involving synthetic (man-made) colloidal particles—especially semiconductors such as cadmium sulfide—was published in the open literature at that time. My group and others modified these methods and developed related ones to prepare colloidal metal nanoparticles, for example, platinum, rhodium, and gold.

In general, the synthesis methods are based on reacting a metal-organic compound in a way that frees the metal atoms from the organic components and causes the metal atoms to coalesce, nucleate, and grow into nanosized metal particles. Chemical suppliers offer a nice selection of these metal-organic "precursors," compounds such as rhodium acetyl acetonate and the platinum analog, that are the forerunners of the products. The reaction is carried out in liquid solution—often an alcohol—in the presence of an organic (polymeric) compound that caps the particles and prevents further aggregation. Although these polymer caps inhibit small particles from agglomerating and fusing together, they nonetheless permit reactant molecules to reach the underlying catalytically active sites and they allow products to escape as they are formed.

With experimentation, we learned to fine-tune the procedure and successfully synthesized a wide range of stable (long shelf life) and uniformly sized metal particles with diameters in the 1-10 nm range. By adjusting reagent concentrations, reactions times and temperatures, and other growth parameters, we were able to tailor the sizes and shapes of these particles. We made microscopic cubes, as well as cuboctahedra and particles shaped like triangles, pentagons, hexagons, and other shapes. We also used these methods to choose whether the metal particles were distributed in two dimensions or three.

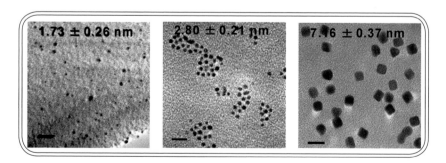

Pt nanoparticles of different size

To make 2D nanoparticle arrays, similar to the ones we made via lithography, we used the well-known Langmuir-Blodgett method. That method orders molecules (or atoms) through compression and transfers them to a solid via a simple film-flotation procedure. The technique was named for General Electric research scientists Irving Langmuir, winner of the 1932 Nobel Prize in Chemistry, and his longtime research colleague, Katharine B. Blodgett. She was the author of the highly cited 1935 research paper describing the thin film technique named for the GE researchers.

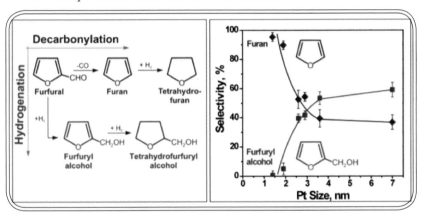

Armed with synthesis tools for finely controlling the size of catalyst particles in the 1-10 nm size range, we showed that product selectivity in some reactions depends strongly on particle size. For example, when furfural reacts with hydrogen in the presence of 1-2 nm platinum particles, the reaction yields furan almost exclusively. Larger particles lead to a mixture of furan and furfuryl alcohol.

Colloidal chemistry provided us with other useful features in addition to particle size and shape control. One of the key ones was the ability to make 3D arrays of particles. The trick here was to distribute the particles throughout the pores and channels of porous oxide materials such as a type of silica known as SBA-15. That arrangement was truly beneficial because it led to high densities of well dispersed nanoparticles, which resulted in very high-surface-area model catalysts that closely resembled industrial catalysts. This type of chemistry also enabled us to prepare multicomponent nanoparticles including bimetallic ones, for example Pt-Pd nanoparticles, as well as particles featuring a core made of one material and a shell made from another one. We also used this type of chemistry to prepare large-pore (mesoporous) metal oxide materials.

All in all, these colloidal chemistry techniques provided us with a very versatile collection of procedures for preparing a large variety of small, uniformly sized and shaped ("monodisperse") nanoparticles—and elegant ways of controlling their composition. After decades of helping to develop the "encyclopedia" of classic single-crystal preparation methods for UHV surface science, we had completely planted ourselves at a new frontier of traditional solution-phase chemical synthesis; complex colloidal nanoparticles.

Having undergone this major shift in experimental approach, we were now armed with adaptable tools for making model catalysts that truly emulated their industrial counterparts—and the catalysis data began pouring in quickly.

We found that some reactions, such as ethylene hydrogenation, do not vary with catalyst size and shape. The switch from macroscopic platinum single crystals to "big" lithographically-prepared Pt nanoparticles caused no significant change in chemistry. Switching to even smaller Pt nanoparticles also caused no change.

Cyclohexene dehydrogenation to benzene, however, is a different story entirely. The turnover frequency, which is a way of expressing the number of molecules that undergo reaction every second—meaning the rate of reaction, is acutely dependent on the size of the platinum particles. So is the activation energy, a measure of the "push" needed to induce reaction. The tiniest particles were the most active catalysts by far.

The thing to appreciate is that our new catalyst preparation methods revealed huge differences in those reaction parameters when the catalyst particle size was varied by just a half nanometer! We ran tests on a half dozen samples (i.e. a half dozen batches of many trillions of identical particles) ranging in size from roughly 1—7 nm. Small differences in the sizes of those tiny particles— even over that narrow size range—strongly affected the particles' catalytic properties.

What about shape? Could the shape of a cluster of metal atoms measuring barely a few billionths of a meter across make any difference in that little particle's ability to mediate a catalytic reaction? Absolutely. And now, for the first time, my research group was able to probe that subtlety.

We investigated that issue by analyzing the hydrogenation products of benzene (C_6H_6), which can include cyclohexene (C_6H_{10}) and cyclohexane (C_6H_{12}), depending on conditions. It turns out that cube-shaped platinum nanoparticles convert the aromatic molecule (benzene) into a single product—cyclohexane. More complex shaped cuboctahedra, on the other hand, convert benzene to a mixture of cyclohexene and cyclohexane. We observed similar results (dependence of product selectivity on particle size and shape) in studies of furan, pyrole, and other compounds important to the chemical industry.

So why would particles of one shape lead to one type of product and some other shape yield other products? The short answer is that it's based on differences in the step-by-step chemical reaction energetics that depend on the various crystal faces exposed by the particles. We came to that conclusion by comparing nanoparticle catalysis data to single-crystal data.

Allow me to elaborate just a bit further. The key concept here is that certain elementary reactions require slightly less energy to proceed on a given crystal face than on others. The less costly reaction (in terms of energy) is nearly always favored. Likewise, when comparing the outcomes of two types of reactions on any single face, again—the less costly reaction will nearly always be favored.

Those concepts apply as well to reactions that could lead to multiple products, but the details can quickly get complicated. In one study of this type, we mapped the product selectivity of hydrogenation of crotonaldehyde to the sizes of the mediating nanocatalysts. Very briefly, that starting material, which is a common reagent in industrial ("fine chemical") synthesis, can be converted to propene (generally undesirable) or to butyraldehyde and crotyl alcohol, each of which can then be converted to butanol. The key finding in this study is that smaller nanoparticles guide the reaction toward butyraldehyde. Larger ones favor crotyl alcohol. Once again, this important distinction was uncovered by comparing the performance of catalytic particles that varied over the incredibly narrow size range of just 1—7 nm.

If it seems puzzling that a difference in particle size of just a few billionths of a meter can make such a large difference in a particle's catalytic properties, consider the following. A metal particle of roughly 1 nm in diameter only contains about 35 atoms. Most of those atoms are exposed—that is, they reside at the surface. They have to be exposed because there is no bulk—no exterior to speak of—in the case of such a tiny speck of matter. As the size changes just slightly—say by 1—2 nanometers, a substantial fraction of the atoms become hidden in the particle's interior. And as the interior of the particle develops, the exterior develops too, forming atomic steps, terraces, and other structural features that often play a central role in catalysis.

By this point in our research we understood just how valuable these colloidal synthesis methods were to catalytic surface chemistry. Tiny differences in the sizes of microscopic particles—as well as differences in their shapes—turned out, sometimes, to be critically important to the rates and the product selectivity of reactions mediated by those particles. And now we had exquisite control over the uniformity of their sizes and shapes as never before.

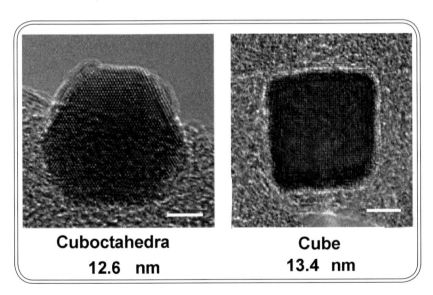

Cuboctahedra
12.6 nm

Cube
13.4 nm

Catalytic nanoparticles, such as the platinum ones shown here, can be prepared in a variety of shapes, including cuboctahedron (left) and cube (right). Studies show that particle shapes strongly influence chemical product selectivity.

In the past, very few chemical reactions—ammonia synthesis among them—were deemed dependent on catalyst particle size. For many other catalytic reactions, that subtlety went unnoticed because common catalyst preparation methods could not ensure the level of microscopic uniformity needed to reveal that subtlety. Some of those reactions were judged to be independent of catalyst structural differences. Now we had the tools to show that some reactions long deemed structure insensitive were indeed highly sensitive to catalyst structure.

The issue of understanding why a chemical reaction might produce more than one product—and how to control the relative abundance of each product—is among the most important scientific challenges of the 21st century. Mastering—or at least favorably manipulating—chemical reaction selectivity, as that topic is known, means being able to steer reactions toward the desired product and inhibiting production of wasteful byproducts.

In essence, that's what "green" chemistry is all about and it is also one of the central quests of catalysis-based surface science. Being labeled "green" or environmentally conscientious for seeking the molecular basis of chemical selectivity is completely appropriate because through that knowledge, we can use energy and natural resources more efficiently than we do presently.

How so? By guiding a reaction toward the desired product, for example by catalyzing the reaction with nanoparticles of tailor-made size and shape, we avoid wasting natural resources and energy making the unwanted products. We also avoid further waste incurred in having to deal with the byproducts. Just to give one example, the "waste" in that case includes the enormous costs and fuel usage associated with distillation and other industrial methods for chemical separations.

This period of research during which we focused on nanoscale catalysts was incredibly rewarding and intellectually stimulating. We learned a great deal about the catalytic properties of these tiny specs of matter. We learned about the importance of their sizes, shapes, and 2D or 3D geometry; about the significance of their composition—be it single element, bimetallic, multimetallic, or oxide coated; and about the influence of their oxidation states and oxide materials on which they are supported.

We also learned about the rapid way in which nanoparticles respond to changes in their chemical environment. Our findings in that area came in part from studies based on high-pressure X-ray photoelectron spectroscopy (XPS). In the previous chapter, I told you about the development and application of a number of analysis methods that enabled us to study catalytic surfaces at relatively high pressure. I briefly mentioned the innovations in XPS, which were led by Miquel Salmeron and colleagues at Berkeley.

The salient point here is that we were able to use this high-pressure variant of XPS to probe nanoparticle behavior; of particular interest was the behavior of bimetallic particles. Specifically, by irradiating a film of these nanoparticles with the intense X-ray beam generated in a synchrotron light source, we were able to distinguish the elemental composition of the particles' tiny bulk and that of their surfaces. We then exposed the bimetallic nanoparticles to a high pressure of an oxidizing gas and later to a reducing gas, all the while monitoring changes in the particles' surface composition.

That investigation showed us how quickly the surface composition of nanoparticles can change as one type of gas preferentially draws one element to the surface and another gas attracts a different metal. Those studies underscored unique properties of nanoparticles and clarified for us the reason that catalysts so often take the form of nanoparticles. The gist of it is nanoparticles are dynamic entities. Their tiny diameters and the reduced coordination of many of their surface atoms enable interior atoms to diffuse quickly across those miniscule lengths. That process leads to rapid environmentally-induced structural and compositional changes that facilitate catalytic reactions.

The nanoscale dimensions of these fascinating particles also provided us with the opportunity to learn about the interface between these microscopic bits of metal and the metal-oxide materials on which they are so often supported. For years researchers knew that the oxide-metal interface plays an important role in controlling catalytic activity and selectivity, even though the oxide by itself usually seemed fairly inert. Studies of nanoparticles helped solve that catalysis contradiction.

It turns out that certain reactions that release energy to their surroundings can do so in a way that generates energetic or "hot" electrons. Normally hot electrons in a solid lose their excess energy through interactions with the solid lattice. This energy dissipation process takes place very quickly—within a few femtoseconds (10^{-15} seconds) before the hot electron can travel more than say 5 nm. But because that length scale is close to the size of these catalytic metal particles, sometimes hot electrons in metal nanoparticles can migrate to the oxide interface at the site of an adsorbed molecule, deposit their energy there, and stimulate a catalytic reaction.

In a similar way, the interface between an adsorbed molecule and an oxide support can be an active host for the type of charge-stimulated reactions that characterize much of acid-base chemistry. Reactions involving charged species—atomic and molecular ions and electrons—often proceed in a way that is distinct from uncharged species. In some cases, the oxide-molecule interface can serve as a catalytic site to facilitate transfer of charge between nanoparticles and nearby molecules in a way that stimulates catalytic reactions.

As my research group and others continued studying nanosized catalyst particles, it became clear that the tiny size of these species endows them with unique properties relative to their bulk-sized counterparts. At the small size of the particles we studied, there is—in actuality—little bulk to speak of, as mentioned earlier. In fact, for some of the smallest particles, nearly all of the atoms reside at the surface. As a result, these heterogeneous nanoparticle catalysts resemble homogeneous catalysts, which generally are described as individual molecules that consist of a metal atom—or several metal atoms—bound to molecular appendages (ligands).

Historically, catalysis researchers have tended to think about these two types of catalysts as distinct entities. And for good reason. The methods, reaction conditions, and applications associated with homogeneous catalysts are rather different from the ones that pertain to heterogeneous catalysts. Homogeneous catalysts, such as the ones used in the polymer industry, are typically used in liquid solvents at fairly mild temperatures to mediate reactions between liquid phase reagents.

In contrast, heterogeneous catalysts, by definition, are not the same state of matter as the reagents between which they facilitate reactions. They are typically solid catalysts used at high temperature and high pressure to drive reactions between gas-phase molecules (with some exceptions). Yet the catalytic centers in both types of catalysts are generally nanometer sized. For example, so-called single-site olefin polymerization catalysts, which consist of a metal atom surrounded by a few other atoms and some organic groups, are often close to two nanometers in size. And of course the nanoparticle catalysts discussed throughout this chapter are nanosized.

The same is true of biology's catalysts—enzymes. Long ago biology figured out how to mediate chemical reactions critical to energy conversion and a host of other key cellular functions in plants and animals through the actions of complex structures known as enzymes. Although the overall structures of these life-sustaining biomolecules are large and exhibit complex 3-D shapes, the catalytic hot spots or active sites tend to be nanosized. The amazing thing about these kinds of catalysts, thousands of which are found in human bodies, is that they do their jobs (in healthy species) with perfect selectivity, in biological fluid (mostly water), and at ordinary plant and animal (body) temperature.

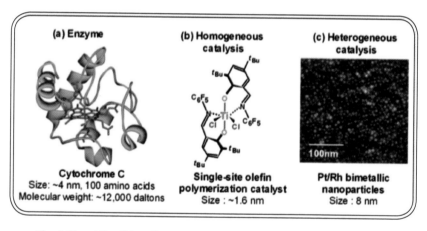

(a) Enzyme	(b) Homogeneous catalysis	(c) Heterogeneous catalysis
Cytochrome C	Single-site olefin	Pt/Rh bimetallic
Size: ~4 nm, 100 amino acids	polymerization catalyst	nanoparticles
Molecular weight: ~12,000 daltons	Size : ~1.6 nm	Size : 8 nm

It's all Nano! Traditionally, enzymes (left), single-molecule homogeneous catalysts (center), and nanoparticulate heterogeneous catalysts (right), seemed radically different from another. In fact, the catalytic hot spots (reactive centers) in all three types are nanometer sized.

Clearly, catalysis is well mediated by nanosized structures and clearly nature prefers highly efficient ones that work well under very mild conditions. With continued advances in understanding the molecular level workings of all three types of catalysts, science should soon be able to develop a unified approach to catalysis that readily shares ideas and concepts between all the areas. By drawing on lessons learned across the catalysis spectrum, this important field of science will be able to uniquely contribute to improving our management of natural resources and energy supplies and to advancing human health and well being.

I have been fortunate to be surrounded here at Berkeley by colleagues who stimulate and inspire me with their seminal contributions to nanoscience and catalysis. Paul Alivisatos, my son-in-law, is a professor of chemistry and materials science and serves as director of Lawrence Berkeley National Laboratory. His highly acclaimed work on synthesis of semiconductor nanoparticles, such as cadmium selenide and cadmium sulfide, helped establish the relationships among these particles' sizes, colors, and electronic structures. He advanced the science of nanoparticle shape, structure, and composition and contributed to our understanding of the particles' optical properties and the behavior of organic molecules adsorbed on them.

Chemistry professor Peidong Yang has excelled at using large numbers of nanoscale building blocks to construct complex architectures with novel electronic and photonic properties. Among his many focus areas, Peidong's work in exploiting light-induced splitting of water to generate hydrogen and oxygen is creative and critically important for advancing the fields of energy conversion and energy storage. Recently we worked together to develop new types of nanocrystal tandem catalysts—nanosized sandwich structures featuring multiple metal–metal oxide interfaces capable of catalyzing sequential reactions. The study describes strategies for designing high-performance, multifunctional nanostructured catalysts.

I have also had the good fortune of working with Dean Toste, a Berkeley chemistry professor and organic chemist, who is an expert in the field of homogeneous catalysis. Guided by Dean's insights, we have been moving forward to bridge the fields of heterogeneous and homogeneous catalysis, by turning to metal nanocluster catalysts in place of traditional homogeneous catalysts that typically feature molecules consisting of single transition-metal ions.

In one example of our studies, we showed that selectivity in heterogeneous catalytic reactions can be controlled by tailoring the properties of dendrimer-encapsulated gold nanoclusters—a strategy reminiscent of modifying homogeneous catalyst ligands. We also demonstrated that product selectivity could be tuned by controlling the reactants' residence time—that is, the time reactants spend in contact with these encapsulated catalysts—inside a type of chemical reactor known as a flow reactor. That investigation brings together some features of homogeneous catalysis with heterogeneous catalysis to improve reaction profiles.

As a result of these studies and related ones, Berkeley has established itself as a leading institution in the discovery, investigation, and development of nanomaterials and their unique properties. I am fortunate to be able to maintain my research program here and proud to be affiliated with such outstanding scientists.

Awards and accolades

The period since 2000 has been a fascinating and highly productive one for my research group. It also is one in which several organizations and institutions graciously honored me for scientific accomplishments. In 2002, I was appointed University Professor of the University of California system, which is the highest academic honor bestowed on a faculty member by the Regents of the University. This honor comes with a responsibility of spending time lecturing at other campuses of the University in addition to my home campus. So I launched a yearly symposium entitled "Applications of Surface Science," which takes the form of a two-day scientific meeting that is convened throughout the campuses of the University of California system.

2002 is also the year in which I received the National Medal of Science. This award is truly special. It is the highest U.S. national honor bestowed upon a scientist—and it is done at a White House ceremony led by the President of the United States.

Our whole family including children and grandchildren attended. When President Bush walked into the room where the ceremony was being held, the silence and decorum was abruptly suspended by my 3-year old grandson, Benjamin, who shouted "That's President Bush!" Benjamin recognized the president from photos—and his enthusiastic outburst instantly lightened the mood of the otherwise dignified yet somber ceremony.

In 2007, I received the Irving Langmuir Prize in Chemical Physics from the American Physical Society. And in 2008, I was honored by the American Chemical Society with the Priestley Medal, which is ACS' most prestigious award. During the same period I received several honorary degrees including those from the University of Manchester and Cardiff University in the U.K., the Swiss Federal Institute of Technology (ETH) in Zurich, Switzerland, the Institute of Chemistry of the Chinese Academy of Sciences, and Northwestern University.

In 2011, I received the Frontiers of Knowledge Award in Basic Sciences from the BBVA Foundation in Spain; the Frontiers of Hydrocarbon Prize from the ENI Foundation in Italy; and the Honda Prize from the Honda Foundation of Japan. Each and every award leads me to reflect on how fortunate I have been to lead a research group filled with so many bright and talented young researchers and to do so at a world class center of scientific research.

CHAPTER 9

A Global Surface Science Network

A colorful collection of traditions, cultures, and accents gives an
international flavor to the web of science contacts we've made
over the years. We are unique yet similar—brought together
by the shared language of surface science and catalysis.

IN THE FINAL 20 years or so of the last millennium, surface science and its applications moved to the frontiers of chemistry. The field grew rapidly and so did my web of international science contacts, friends, and collaborators. Allow me just a brief digression from the stories I have been recounting about the development of the *science* part of surface science to tell you a little about some of the important *people* behind the science. In particular, let me briefly mention some of the people who helped expand our global surface science network in the period from roughly 1980—2000.

I had the good fortune to work with some wonderful Japanese scientists including Yoshimasa Nihei, who was a postdoctoral fellow in my laboratory. He later went on to become a professor at the University of Tokyo. Yasuhiro Iwasawa, a University of Tokyo chemistry professor and a surface science and catalysis specialist, also spent time working in Berkeley. I have remained in contact and stayed friends with both of these researchers.

Among my colleagues from France, Jacques Oudar and Christian Minot, professors at the University of Pierre and Marie Curie, Paris, and Professors Gault and Goldstaub in Strasbourg all stayed in working contact with me over the years and sent several of their graduates to me to work as postdoctoral fellows.

I also developed lasting contacts in the United Kingdom including Jack Linnett, who served as Head of the Department of Physical Chemistry at the University of Cambridge; Ron Mason, a professor at the University of Sussex; and Wyn Roberts, who was a professor at the University of Bradford and later at Cardiff University in Wales. I also recall fondly my interactions with John Pendry, a professor at Imperial College London, and with Professor David King, chancellor of the University of Liverpool. Over the years, I also had the opportunity to work with many excellent postdoctoral scientists from the U. K.

I've also had the opportunity to develop lasting relationships with scientists in Sweden and neighboring countries. For example, I became good friends with notable surface science and catalysis experimentalists and theoreticians, including Ragnar Larsson at Lund University, and Ulf Karlsson, Anders Rosengren, and Christofer Leygraf of KTH Royal Institute of Technology in Stockholm. Christofer spent time at Berkeley as a postdoctoral fellow.

Over the years, I befriended several surface scientists and catalysis researchers in Finland. They include Outi Krause, who has held several academic positions in Finland and was recently appointed dean of Aalto University's School of Chemical Technology. Also included are Riita Keiski at the University of Oulu and Jouko Lahtinen in Aalto University's Department of Applied Physics. For years I was a member of the International Advisory Board of Neste Oy, an oil and energy company in Finland.

Our research group made plenty of science connections with other countries in Europe. In Austria, for example, we developed a close working relationship with Konrad Hayek, a professor of physical chemistry at the University of Innsbruck. And Konrad's PhD student, Günther Rupprechter, worked in my lab as a postdoctoral researcher and later became a professor of physical chemistry at Vienna University of Technology.

In Italy, Paolo Galli, who served for years as Research Director of Montecatini, a large chemical manufacturer in Ferrara, and later took a position in academia, was a dear colleague of mine. He encouraged several of his students to conduct postdoctoral research with me in Berkeley. Paolo and I worked together in the field of catalytic polymerization of propylene.

Elsewhere in Italy, Massimo Simonetta, a professor of physical chemistry at the University of Milan, was among the first people to notice the low energy electron diffraction work we did and recognize its importance in determining the surface structures of clean crystals as well as those covered with adsorbed molecules. Simonetta had spent time as a visiting scholar at Caltech working with Linus Pauling deducing the structures of molecules from X-ray studies. With that background, he was keenly interested in molecular structures and worked to reveal "secrets" of chemical bonds.

I have long enjoyed a close and important relationship with scientists from Spain. One of my collaborators was Juan F. Garcia de la Banda, director of Madrid's catalysis institute. But it was Nicolas Cabrera, a physics professor at Autonomous University of Madrid who had a major impact on the direction of surface chemistry in Berkeley. His student, Miquel Salmeron, joined my group as a postdoctoral fellow in the 1970s and remained in Berkeley as a staff scientist at Lawrence Berkeley National Laboratory.

Miquel is an invaluable collaborator and an outstanding scientist. We have been working together closely for several decades and have succeeded in designing novel instruments and techniques for uncovering new phenomena in surface science. Miquel's prestige has attracted a handful of other postdoctoral fellows from Madrid and Barcelona, including Salvador Ferrer, Felix Ynduráin and Juan M. Rojo. These researchers have added a delightful Spanish flavor to surface science research in Berkeley.

Among colleagues from the Netherlands, I warmly acknowledge my wonderful relationships with Rutger van Santen and Hans Niemantsverdriet, a well-known catalysis theoretician and surface science experimentalist, respectively, at Eindhoven University of Technology. Both researchers spent time at Berkeley on sabbatical leave. Also from the Netherlands, were Wolfgang Sachtler, who spent much of his career at Northwestern University, and Vladimir Ponec. Both scientists sent outstanding graduates (Adrian Sachtler and Gerard Vurens) to work with me as postdoctoral scientists.

I have had the chance to work with several delightful postdoctoral fellows from Belgium. One of them, Marie-Paule Delplancke, is connected to me on several levels. In addition to working with me as a postdoc, she married a student in my group, Frank Ogletree, and she also translated the first edition of my textbook, *Introduction to Surface Chemistry and Catalysis*, into French.

There was a time in the 1980s when I briefly explored the possibility of moving back to Europe. Here is what happened. Piero Pino, an internationally acclaimed scientist in the fields of polymerization and homogeneous catalysis, was visiting me in Berkeley in 1987. At that time, he was nearing mandatory retirement age and he proposed that I consider succeeding him as a research director at the ETH in Zürich, Switzerland, a world-class academic institution.

I was greatly honored by the offer and, although I was very happy in Berkeley, I decided to take a three month visit to his laboratory to experience life at the ETH up close and in person. As it turned out, Swiss salaries are very generous even by U.S. standards. But an even stronger attraction for me was the substantial financial support for research equipment, and additional support in the form of permanent staff scientists and postdoctoral fellows. All of that support was part of a package offered to new professors there.

Professor Pino was a top-notch scientist, a kind person, and a dear friend. We got along very well. At the end of the three month visit, after consulting with my family, we decided that the United States is our home and we just could not see resettling in Europe after our experiences in the New World. So I gratefully declined the offer.

But that's not the end of the story. I was able to make a valuable recommendation to the ETH. It so happened that Roel Prins, an outstanding catalysis scientist and expert in X-ray absorption spectroscopy from Eindhoven University of Technology, was on sabbatical leave at that time working with me at Berkeley. So I asked Roel whether he would be interested in the position at the

ETH, and indeed, he was very interested. In the end, Roel moved to Zurich and filled Professor Pino's position with great success. Ever since that time, the ETH administration has trusted my judgment. For example, when they were looking to fill a faculty position in the newly formed Department of Materials Sciences, they accepted my recommendation to name Nicholas Spencer to that position. Nic had previously worked with me as a postdoctoral researcher and he has since launched a spectacular career at the ETH and has developed a reputation as a tribology expert.

Around the same time the previous story was unfolding, I was invited to deliver a presentation at a photochemistry conference in Budapest, Hungary, the city in which Judy and I were born. As it happens, I was working at that time on the photodissociation (light-induced breakup) of water to hydrogen and oxygen—and the meeting was nicely aligned with some of my research interests. So I decided to go.

Now, Judy and I had not been back to Hungary since we had escaped in 1956 following the crushed Hungarian Revolution. And because of our escape, Hungary had sentenced us, *in absentia*, to 10 years of hard labor. The Hungarian government eventually granted general amnesty to Hungary's escapees, but it was never very clear just how safe we would be under Hungarian and international law, even as naturalized American citizens, if we chose to return to our birth country. But now we felt confident enough to take the trip.

So in 1987, for the first time in 31 years, we returned to Hungary. We were welcomed warmly. I received an honorary doctorate from the Technical University of Budapest, where I had been an undergraduate student at the time of our escape, and an honorary degree from the University of Szeged. I was also appointed as a Foreign Member of the Hungarian Academy of Sciences.

Ever since that visit, we have continued to return to Hungary every few years to show our birthplace to our children and grandchildren. In addition to taking family trips, I have tried to help Hungarian science in some ways, for example, by finding positions for young and senior Hungarian scientists in my laboratory and by participating in collaborative research projects.

I would be remiss if I didn't name a few of the Hungarian scientists with whom I have had the pleasure of working and interacting over the years. The late Professor Ferenc Márta, the one who invited me to the 1987 photochemistry conference, was a kineticist and Director of the Chemical Institute of the Hungarian Academy of Sciences. I developed a close working relationship with a number of scientists at the Hungarian Academy—especially Gabor Palinkas, who succeeded Márta as director. I chaired the International Advisory Board of this institute, and in that role, developed lasting contacts with several scientists working in a variety of fields throughout the chemical sciences.

One such person with a Hungarian family history is Peter Stang, a well-known organic chemistry professor at the University of Utah and editor in chief of the Journal of the American Chemical Society. I have also had the distinct pleasure and knowing and enjoying the friendship of another Hungarian expatriate—Professor George Olah, a chemistry Nobel Laureate and Director of the Loker Hydrocarbon Research Institute at the University of Southern California.

In 1989 I spent six months in Berlin (at that time, West Berlin) as a Humboldt Foundation Senior Fellow at the Fritz Haber Institute. Over the years, I have remained in close contact with several leading scientists at FHI including Hajo Freund and Robert Schlögl. While I was visiting in '89, my host was chemistry Nobel Laureate Gerhard Ertl.

Naturally, spending time in that famous institution led me to reflect on its history and to think about the brilliant scientists who established its reputation.

In the 1920s, the laboratory was home to influential physical chemists Fritz Haber and Michael Polanyi. Haber, after whom the institute was eventually named, was a German chemistry Nobel Laureate who, to this day, remains famous for his seminal work on ammonia synthesis. Polanyi was a Hungarian scientist, who is remembered for his work on chemical reaction mechanisms.

Among other reactions, Polanyi studied catalytic hydrogenation and hydrogen-exchange in ethylene—reactions I also studied. But unlike my research group, which had a host of modern surface science tools for spectroscopy, microscopy, and surface diffraction studies, Polanyi had very few tools in the early 1900s with which to study those reactions. Yet his keen powers of observation and chemical reasoning—and a set of data from isotope-exchange reactions were sufficient to enable Polanyi to propose the elementary steps that comprise reaction mechanisms. Clearly, Michael Polanyi, an unsung hero who helped develop chemical reaction dynamics and kinetics, was a man with deep chemistry insights.

The other bit of history that came to mind while I was in Berlin was the fate of these scientists after Hitler rose to power. Regardless of their religious practices and beliefs, Haber and Polanyi both lost or resigned from their academic positions because they were born to Jewish families. Polanyi moved to England and continued his scientific career there. Haber accepted a position at the research center that would later come to known as the Weizmann Institute in Israel, but he died en route to the Middle East. It was interesting to reflect on the fate of these scientists, both of whom contributed substantially to chemistry.

Soon after Judy and I arrived in Berlin, we spent an evening enjoying dinner at the home of Barbara and Gerhard Ertl. During dinner I commented that the upbeat mood in Berlin reminds me of what we experienced in Budapest in 1956, just before the Hungarian Revolution. Just as Hungarians at that time were openly critical of the government and less concerned with repercussions of taking liberties than they had been just a year earlier, Berliners seemed to be acting in much the same way. At that time, Communist officials from East Berlin freely shopped in the upscale West Berlin department store KaDeWe—and though they were clearly violating Communist restrictions, no one seemed to care. "You never hear anymore that people are shot trying to escape from East Berlin," I told the Ertls. The once-tough Communist regime seemed to have become more tolerant.

"The bullet is still stronger than the brain," Ertl replied. People's fear of being terrorized by the government will continue to subside— and eventually Communism will collapse, he predicted, but not for another 20 years. I didn't think he was right. "I'll bet you a bottle of champagne we'll witness major changes within a year," I told him. "Why just one bottle? Let's make a case of champagne," Barbara suggested. We agreed.

In November of 1989 the Berlin Wall was demolished. In the following year at the American Chemical Society meeting in San Francisco there was a symposium organized to celebrate my 55th birthday. One very happy Gerhard Ertl arrived at the meeting with a case of champagne for me and a cement block souvenir from the Berlin Wall. He was delighted to have lost the bet!

Let me mention just one last country—China. In 1982, the United Nations Educational, Scientific, and Cultural Organization (UNESCO) sent a delegation of chemists including me to China on a mission to help advise Chinese educators on starting Ph.D. programs in chemistry. Effectively, there were no such programs in China at that time.

We were welcomed warmly at Fudan University in Shanghai, where I met the university president, Professor Xie Xide, and her husband, who at that time headed of one of the bio-institutes of the Chinese Academy of Sciences. We visited Xian, where an army of clay soldiers had been unearthed just a few years earlier, and also visited Beijing and Changchun.

In Changchun, we met with scientists at Jilin University. Our delegation identified a number of key English textbooks as important teaching materials for university students—and recommended making Chinese translations widely available. Jilin University, where I spent two weeks, was responsible for organizing the translations. In addition to recommending curriculum improvements and other types of changes, we also suggested that Chinese institutions evaluate the effectiveness of their education system and student preparedness—in what today would be called a "transparent" manner. Our suggestion was that students be required to take Graduate Record Examinations. GREs are standardized tests that are typically required for admission to U.S. graduate schools.

It was clear to me and to our delegation that China was serious about ramping up its science education nationwide. What was not at all obvious at that time was just how quickly these changes would be implemented and how rapidly the level of Chinese science education—and the technology applications that tend to follow from advanced education—would blossom. In the early 1980s, China was sitting on the cusp of major progress in science and technology. In the short intervening time, the country's scientists and engineers quickly moved the center of the world stage in those areas.

Many students now obtain a top-notch graduate chemistry education in China, while others still choose to come to the U.S. for graduate and postdoctoral work. The U.S. in general, and my research group in particular, have greatly benefited from the research efforts of this pool of talented, hard-working young scientists.

Let me wrap up this section on the global network of science contacts we have developed with a word about gratitude—an emotion that isn't limited by country. Some years ago, Judy and I wanted to show our appreciation for the wonderful opportunities that have come our way through a life of science, and in the process, help other scientists. So we decided to establish endowment funds that support awards for scientific achievement.

One of the awards is the Gabor A. and Judith K. Somorjai Visiting Miller Professorship Award. The aim of the award is bring promising or eminent scientists to the Berkeley campus on a short-term basis to carry out collaborative research. The other award is the American Chemical Society Somorjai Award for Creative Research in Catalysis Science. We gratefully acknowledge the encouragement and support we received from ACS Executive Director & CEO, Madeleine Jacobs, in establishing this prize. The award was set up to recognize outstanding theoretical, experimental, or developmental research resulting in the advancement of understanding or application of catalysis.

CHAPTER 10

Closing Thoughts

A Historical and Personal Perspective on Science in the Last Fifty Years of the 20th Century & Beyond

I HAVE LAID OUT my life story for you over the past several chapters. Now let me try to wrap up by distilling my thoughts down to a brief personal perspective on the period in which I lived.

Most of my working years spanned the second half of the 20th century. In that period, there was a spectacular rise in the world's collective understanding of science. The knowledge gained during that time frame made a powerful impact on the health, wealth, and quality of life in the United States and around the globe. I am proud to say that surface science played a role in those advances.

The field of science I worked in progressed over the years toward ever finer analyses and more detailed descriptions of surfaces and interfaces. This body of knowledge, with its treasure trove of atomic- and molecular-level subtleties, underpins modern science's understanding of numerous phenomena in catalysis, electronics, optics, and materials science. And it has shaped our understanding of biointerfaces, polymers, coatings, as well as mechanical properties of materials, such as friction, lubrication, corrosion, and wear.

These accomplishments in fundamental science have spawned new technologies in many vitally important fields. They include energy conversion processes for efficiently tapping oil and gas supplies, solar power, and other renewable sources of energy. They include advances in batteries and other aspects of energy storage, as well as the emerging area of biomass conversion.

These important steps forward in basic science have also led to valuable advances in health care. Examples in that domain include development of new medical instruments and associated methods for non-invasive investigation of the human body, as well as tools and techniques for repairing the human heart, brain, various other organs, and bones. These discoveries have also led development of new drugs that have helped increase our life expectancy and vastly improved our quality of life.

In a similar way, this progress in science has helped bring about key advances in agricultural technologies and food production. It has also led to environmental benefits, including improved air and water quality throughout much of the world and new technologies for environmental remediation.

Advances in surface science have also had a direct impact on various areas of electronics including microprocessors, data storage and transmission hardware, and other components of modern high-power computing. Those developments, in turn, spawned a revolution in communication and information technology, which has enabled a massive flow of data, knowledge, and novel ideas for the education and benefit of many millions of people.

As I think back on my years in academia, I am reminded about my family and my beginnings. My parents came to the U.S. in 1962 after I was naturalized as a U.S. citizen and they settled in Berkeley where I was appointed to the chemistry faculty in 1964. They opened a small children's clothing store and supported themselves modestly by working there together until my father passed away in 1983.

My father was a very proud man. He never agreed to accept any financial support from me. He was endowed with considerable creative talents but was compelled to use his talents and energy just to survive and to help his family under difficult circumstances.

I cannot help but reflect on the irony of life that gave me the opportunity to do new science and to put my stamp on the future of science by educating generations of creative students. I was blessed to be able to use my talents and energy for these constructive purposes. My father unfortunately was obligated to spend all his talents and creativity on making ends meet, staying alive, and helping his family succeed.

My life's luck is to have survived the Second World War and to have obtained a good education in Budapest in my high school years. The 1956 Hungarian Revolution that erupted and was quickly crushed during my university years gave me, in a roundabout way, the opportunity to immigrate to the United States with my beloved Judy. That tumultuous process landed us in Berkeley, where I was able to continue my education in one of the leading universities and top chemistry departments in the country and, indeed, the world.

As I discussed in earlier chapters, there were several Hungarian immigrant scientists who became leaders in the United States during the first fifty years of the 20th century. Eugene Wigner, Edward Teller, and Leo Szilard were among the physicists in the age of the atom, the atomic bomb, and the development of quantum mechanics. Their worked helped advance our understanding of the structure of matter and unlocked the energy of the atom.

Other Hungarian scientists including chemistry Nobel Laureate George Olah, University of Utah's Peter Stang, and I were among the immigrant flux of the second fifty years of the 20th century. We focused on chemical bonding and chemical structure, and sought to understand the properties of molecules, including large ones such as polymers and DNA. We also strived to uncover the properties of catalytic solids and the inner workings of transistors. These pursuits helped uncover the structural complexity of molecules and led to discoveries in chemistry that have vastly improved the quality of life.

Another thought that loops through my mind as I think back over my long career is the role of U.S. funding in jump starting advanced scientific research. United States government investment in scientific research since the fall of 1957, when the Soviet Union launched the space age by sending *Sputnik I* into orbit, has followed an exponential trajectory. In large part, the goal has been to promote scientific and technological discoveries with the potential to benefit the welfare, economy, health, and general stability of the U.S.

It is clear that the level of federal funding for scientific research at American universities and research institutions in the past half century has been a real game changer in the world of academic scientific investigation. That investment stimulated cutting edge discovery and innovation and made the U.S. the global science and technology leader.

A few funding history statistics, easily found on official funding agency web sites, show how deep research support runs in the U.S.

and how quickly it climbed. At the National Science Foundation, for example, the U.S. Congress responded to the *Sputnik I* launch in 1957 by appropriating $134 million for the newly formed agency. NSF was launched just a few years earlier with a modest budget of $3.5 million. By 2010 the value had soared to roughly $7 billion. A similar trend, but at a different scale, swelled the budget of the National Institutes of Health over the same period of time from roughly $50 million to $30 billion.

The overall level of federal R&D spending, including defense and non-defense-related research skyrocketed from roughly $8 billion in 1950 to something on the order of $150 billion in 2010. During the same time frame, U.S. spending on non-defense-related research climbed from about $1 billion to some $60 billion. The largest allocation increases at the end of that period were awarded to the National Institutes of Health, the Department of Energy, and the National Science Foundation.

This financial support has enabled American universities and research institutions to thrive in science and technology. It has helped train talented and creative young students, inspired them to pursue important research goals, and helped them grow and mature scientifically in a culture that rewards hard work, original thought, and innovation.

In some cases, this culture has motivated young scientists to pursue the opportunity to capitalize on a technological advance by launching a small startup company. The payoff can be substantial, but the risks are high and the chance of failure is great. Yet in the U.S. high-tech startup culture, failing to float a small company financially is not regarded as failing scientifically. Aspiring scientific entrepreneurs, can pick themselves up, get back into science research, and if so inspired, try again to launch a company.

Another outcome of the substantial government investment and support of academic research over the past half century has been a shift in the roles of academia and industry in scientific discovery. During this period, universities, for the most part, took over the role as the principal centers of fundamental or basic science research and discovery. As a result, universities such as UC Berkeley attracted large numbers of bright and gifted students, and I was fortunate to have been able to educate and work with many of them.

In all, I mentored more than 400 students and postdoctoral fellows, teaching them by example ways to develop strategies for research, as well as for organization, focus, and the discipline necessary to complete projects. As a mentor, I have shared with my students my long-term vision of scientific progress in multiple directions and taught them the importance of aiming high and dreaming big dreams. I taught them the importance of mastering material synthesis and instrumental analysis techniques—keys to facilitating scientific discovery. And I also taught of them about the hard work, dedication, and attention to details required for success in science.

Students tend to leave after four or five years to take up new research positions or start their professional careers in science—and new students come to take their places. That ongoing cycle drives the dissemination of new knowledge that stands poised to make an impact on science and technology. This thriving turnover of science and scientists generates a lasting record of scholarly research papers, comprehensive review articles, books, and more recently, videos and other types of science repositories that catalog today's accomplishments for tomorrow's young scientists.

Scientific advances in the last fifty years of the 20th century and the early years of the 21st century led to an explosion of new technology, high-tech products, and processes in so many areas. The entire enterprise improved the quality of life, increased life expectancy, and broadly enhanced communication and education thereby improving the collective human experience and strengthening democratic governance worldwide.

The future of mankind is brighter than it has ever been before. For me, it has been a terrific journey.

ACKNOWLEDGEMENTS

I have been blessed with the loving support of my family and I am forever grateful to my wife, Judy; our children, Nicole and John; and our grandchildren, Stephanie, Clara, Benjamin, and Diana.

It has been my honor and great fortune to serve as a professor in the chemistry department at the University of California, Berkeley, and as a faculty scientist at the Lawrence Berkeley National Laboratory. These institutions are world class. They have enabled me to work with outstanding colleagues and to build and conduct research with sophisticated analytical instruments for over five decades of my professional life. Berkeley is an exciting place to live and work. It is an intellectual powerhouse and a center of scientific and technological excellence that is second to none.

Figure 38: Judy and Gabor with children and grandchildren in Hawaii, December 2012. From left: John, Hilary, Stephanie, Diana, Clara, Paul, Nicole, Benjamin, Judy and Gabor.

For decades, I have been the fortunate recipient of numerous research grants. And I am duly thankful to the United States funding agencies, especially the Basic Energy Sciences program of the Department of Energy Office of Science, and other organizations for making these funds available to support my research program.

Inger Coble, my assistant, has run my office single-handedly for many years. She has been an invaluable support for me and my ever-changing research group. She has my admiration and gratitude.

In my years at Berkeley, I have had the privilege to mentor some 400 talented PhD students and postdoctoral fellows. Their hard work and dedication permitted us to jointly discover new phenomena in molecular level science. I have always told my students to remember that they don't work *for* me. We work together, I remind them, but in truth, you work for yourself. I am here to point you in the right direction and offer suggestions, I tell them. But the knowledge and skills you acquire in my laboratory, you take with you as you leave to forge your careers.

My love of research has rubbed off on many of my students and I have enjoyed watching that process unfold. I have encouraged them to dream big dreams and to choose important and challenging scientific problems that are worthy of working on, even if it takes a lifetime. More than 100 of my students hold leading academic positions and a large number of those who did not pursue careers in academia have become successful professionals in other areas. Their ongoing contributions to science and technology amplify my contributions to surface science and catalysis. Their accomplishments fill me with pride knowing that together we helped advance these fields to the frontiers of chemical sciences.

I have listed here my group members along with our research highlights and seminal discoveries. They are organized by scientific topic.

Somorjai Coworkers 1964–2012

1965	J. E. Lester
	S. Hagstrom
	H. B. Lyon
	Whalen Szeto

1967	A. M. Mattera
	R. M. Goodman
1968	H. H. Farrell
	J. M. Morabito
	A. E. Morgan
1969	R. F. Steiger
	R. M. Muller
1970	T. M. French
1971	F. J. Szalkowski
	L. A. West
	R. Kaplan
	J. G. Davy
	C. Y. Lou
	D. L Howlett
	B. Lang
	M. R. Martin
1972	R. W. Joyner
1973	J. L. Gland
	S. L. Bernasek
	W. J. Siekhaus
	S. Berglund

1974	K. Baron
	D. W. Blakely
	S. B. Brumbach
1975	S. H. Overbury
	L. L. Kesmodel
	M. A. Chesters
	C. Magerle
1976	D. I. Hagen
	T. N. Taylor
	L. E. Firment
	S. T. Ceyer
	R. J. Gale
	J. C. Buchholz
	B. E. Nieuwenhuys
	G. Rovida
	P. C. Stair
	R. C. Baetzold
	Y. W. Chung
	B. A. Sexton
	E. I. Kozak
	Ch. Steinbruchel
1977	W. J. Lo
	A. Jablonski
	M. Salmeron
	R. Bastasz
	J. C. Hemminger

1978	D. Castner
	D. J. Dwyer
	L. L. Antanososka
	K. Yoshida
	L. H. Dubois
	R. Carr
1979	I. Toyoshima
	J. P. Biberian
	C. E. Smith
	H. Ibach
1980	S. M. Davis
	M. A. Van Hove
	F. T. Wagner
	S. Ferrer
	R. L. Blackadar
	P. K. Hansma
1981	I. Bartos
	R. J. Koestner
	W. R. Guthrie
	J. D. Sokol
	R. K. Herz
	W. D. Gillespie
	T. H. Lin
	Y. Nihei
	J. W. A. Sachtler
	H. Heineman

1982	A. L. Cabrera
	P. R. Watson
	N. O. Spencer
	R. C. Schoomaker
	J. M. Rojo
	J. C. Frost
	M. Quinlan
	J. E. Crowell
	C. A. McLean
	M. Langell
	E. L. Garfunkel
	F. Zaera
	P. W. Davies
	C. Minot
	G. Gavezzotti
	M. Simonetta
	M. Asscher
	C. H. Leygraf
	M. Hendewerk
1983	M. H. Farias
	F. Delannay
	W. T. Tysoe
	B. E. Koel
	M. Jazzar
1984	U. Bardi
	C. M. Mate

J. E. Turner

A. J. Gellman

B. E. Bent

J. J. Maj

1985   C. Zhang

J. B. Pendry

M. Logan

K. Sieber

C. Sanchez

E. Pollak

M. M. Khan

M. Kudo

C. H. Mak

K. B. Lewis

A. Hubbard

J. Carrazza

G F. Wang

1986   B. M. Naasz

D. F. Ogletree

D. Godbey

M. R. Hilton

T. G. Rucker

T. M. Gentle

X. Youchang

S. R. Bare

D. R. Strongin

D. G. Kelly

1987	M. M. Khader
	G. H. Vurens
	M. E. Bussell
	G. S. Blackman
	M. I. Ban
	H. Ohtani
	M. E. Levin
	C. T. Kao
	J. A. Odriozola
	G. J. Vandentop
1988	K. J. Williams
	B. Marchon
	S. T. Oyama
	A. L. Slavin
	I. Böszörmenyi
1989	R. Nix
	P. Pereira
	F. Garin
	S. W. Wang
	R. F. Lin
1990	C. Ocal
	P. J. Rous
	M. Kawasaki
	R. D. Levine
	A. B. Boffa

J. Lahtinen

S. H. Lee

S. Fu

R. Q. Hwang

V. Maurice

X. Z. Zhang

1991 P. A. Nascente

J. M. Powers

M. F. V. Van Tol

F. Luck

D. M. Zeglinski

M. P. Delplancke

M. Kawasaki

K. Kobayashi

J. Rasko

A. Wander

G. Held

B. McIntyre

T. Nakayama

K. Takeuchi

T. Katona

L. Guczi

D. Jentz

1992 C. Kim

K. Heinz

U. Starke

C. Knight

N. Materer

1992   D. M. Ginter

E. Magni

D. Gardin

1993   D. E. Wilk

G. D. Stanners

J. Dunphy

Y. F. Chang

O. Desponds

R. L. Keiski

M. Nakazawa

A. Barbieri

P. D. Ditlevsen

W. Weiss

J. D. Batteas

S. S. Perry

H. C. Galloway

J. W. Ager

G. J. Wang

1994   T. Anraku

G. G. Jernigan

F. H. Ribeiro

A. Bonivardi

P. Sautet

X. D. Xie

J. Lahtinen

C. Lin

1995   P. Cremer

R. Döll

J. W. Niemantsverdriet

G.-J. Kroes

S. P. McGinnis

S. Rizzi

U. Schröder

E. K. Starkey

1996   R. L. White

X. Su

P. W. Jacobs

W. J. Wind

D. H. Fairbrother

1997   J. Jensen

C. A. Gerken

A. E. Schach von Wittenau

C. H. Bartholomew

H. A. Yoon

J. G. Roberts

M. X. Yang

G. Rupprechter

D. Zhang

R. S. Ward

A. Eppler

M. Gierer

J. Cerda

J. Wen

L. Lim

1998 L. Lianos

K. B. Rider

D. H. Gracias

Y. H. Park

T. Fujikawa

M. Gaukler

P. D. Cernota

K. Kung

1999 Z. Chen

T. Koranyi

Y. Borodko

S. Hoffer

S. Baldelli

2000 L. Romm

A. Avoyan

S. Niederberger

A. L. D. Ramos

P. Chen

J. H. Song

J. Zhu

S. H. Kim

K. McCrea

G. R. Tewell

2001   G. Mailhot

B. Mailhot

A. Opdahl

D. Ghani

K. Komvopoulos

K. S. Hwang

2002   E. Amitay-Sadovski

C. Marmo

F. Malizia

S. Westerberg

J. Grunes

R. A. Phillips

T. Tsirlin

Z. Chen

K. C. Chou

R. J. Gartside

F. M. Dautzenberg

Z. Konya

V. F. Puntes

I. Kiricsi

A. P. Alivisatos

J. S. Gaughan (Parker)

E. A. Anderson

2003   J. Kim (Joonyeong)

M. Yang (Minchul)

H. Frei

D. C. Tang

Y.-K. Choi

J. Bokor

2004   R. M. Rioux

Chen Wang

T. S. Koffas

C. C. Lawrence (Christopher)

2005   H. Song (Hyunjoon)

J. D. Hoefelmeyer

S.J. Kweskin

K. Niesz

T. D. Tilley

A. M. Contreras

X. Z. Ji (Xiaozhong)

X.-M. Yan (Xiaoming)

A. Marsh

M. M. Koebel

K. A. Becraft

M. Grass

K. M. Bratlie

L. D. Flore

2006	M. Montano
	O. Mermut
	D. C. Phillips
	R. L. York
	J. Y. Park
	C. J. Kliewer
	D. Butcher
	Y. Zhang (Yawen)
	F. Tao
	T. Zhang (Tianfu)
	J. R. Renzas
	B. Hsu (Bryan)
	Y. Li (Yimin
2008	W. K. Browne (Will)
	J. N. Kuhn
	W. Huang (Wenyu)
	L. C. Jones (Louis)
	M. Bieri
	C.-K. Tsung (Frank)
	H. Bluhm
	G. J. Holinga
	L. Belau
2009	S. H. Joo
	C. Aliaga
	H. S. Lee (Hyun Sook)
	H. Seo (Hyuntak)
	A. Hervier

2011 S. Alayoglu

L. R. Baker

R. M. Onorato

C. M. Thompson

F. Zheng (Fan)

V. Pushkarev

G. Kennedy

S. Beaumont

Z. Zhu (Zhongwei)

Y. Wang (Yihai)

2012 J. Krier

E. Gross

L. Heinke

K. An (Kwangjin)

W. D. Michalak

Q. Cai (Rebecca)

N. Musselwhite

A. Lindeman

G. Melaet

H. Wang (Hailiang)

A. Sapi

K. Na (Kyungsu)

Q. Zhang (Joe)

L. Carl

K. Oba

W. Ralston

Fei-Fei Shi

FIELDS OF RESEARCH

A. **Techniques Based On Metal Single Crystals**

    1. Low Energy Electron Diffraction

    2. Molecular Beam Scattering

    3. Low Pressure-High Pressure Methods for combined Surface and Catalysis Studies

    4. Auger Electron Spectroscopy

    5. High Resolution Electron Energy Loss Spectroscopy

    6. High Pressure Scanning Tunneling Microscopy

    7. Sum Frequency Generation Surface Vibrational Spectroscopy

B. **Techniques Based On Nanoparticles**

    1. Sum Frequency Generation Surface Vibrational Spectroscopy

    2. Ambient Pressure X-ray Photoelectron Spectroscopy

    3. Extended X-Ray Absorption Spectroscopy

    4. Extended X-ray Absorption Fine Structure Spectroscopy

C. **Nanoparticle Synthesis**

D. **Catalysis on Single Crystals**

E. **Nanocatalysis and Catalysis on Nanoparticles**

F. **Polymer Surfaces and Biointerfaces**

G. **Hot Electron Catalysis**

H. **Bridging Heterogeneous and Homogeneous Catalysis**

I. **Other Areas of Research**

Seminal Scientific Contributions of my Coworkers, Students and Postdoctoral Fellows

A1. Our low energy electron diffraction studies in ultrahigh vacuum revealed the reconstruction of metal single crystal surfaces of platinum (1965, H. B. Lyon, S. Hagstrom)

A1. Low energy electron diffraction reveals ordered organic monolayers on single crystal platinum (1969, A. E. Morgan, H. H. Farrell, R. M. Goodman).

A2. Construction of molecular beam scattering apparatus from single crystal surfaces under UHV conditions (1971, L. A. West, E. I. Kozak).

A1.	Discovery of atomic steps and other defects that mediate chemical reactions (bond breaking) of adsorbed molecules on platinum surfaces (1972, B. Lang, R. W. Joyner).

A2.	Molecular beam scattering study of hydrogen–deuterium exchange on stepped surfaces of platinum (1973, S. L. Bernasek).

A1.	Adsorbed hydrocarbon molecules react on stepped surfaces, but remain molecularly adsorbed and unreactive on flat platinum crystal surfaces (1974, D. Blakely, J. L. Gland).

D1.	First catalytic reaction: hydrogenolysis of cyclopropane on stepped platinum single crystal at atmospheric pressure (1974, D. R. Kahn).

A1.	First metal surface structure determination by surface crystallography, the (111) crystal face of platinum (1975, L. L. Kesmodel, P. C. Stair).

A1.	LEED studies of the surface structures of vapor grown molecular crystals. Ice and naphthalene (1975, L. E. Firment).

A4.	Auger electron spectroscopy of the composition of alloy surfaces (1975, S. H. Overbury).

A3.	Development of novel instrumentation to monitor catalytic surface reaction over wide pressure range (1976, D. W. Blakely, B. A. Sexton, E. I. Kozak).

A1.	Surface structures of adsorbed ethylene and acetylene on the flat hexagonal platinum crystal face. Evidence of ethylidyne formation (1977, P. C. Stair, L. H. Dubois).

A2. Surface scattering and H-H bond breaking by molecular beams on platinum crystal surfaces (1977, S. T. Ceyer, M. Salmeron).

A1. Chemisorption of molecules on rhodium and iridium crystal surfaces (1978, B. E. Nieuwenhuys, D. Castner).

I. Photon–assisted reactions of water and carbon dioxide adsorbed on $SrTiO_x$ crystal face form methane (1978, J. H. Hemminger, R. Carr).

D. Hydrogenation of CO on rhodium and iron (1979, D. Dwyer, P. R. Watson)

D. Hydrocarbon reactions catalyzed by gold on platinum and clean platinum crystal surfaces (1980, S. M. Davis, W. D. Gillespie, J. W. A. Sachtler).

D. Structure sensitivity of iron single crystals in catalytic synthesis of ammonia (1981, N. D. Spencer).

A1. Dynamical LEED theories developed for large adsorbed molecules (1982 M. A. Van Hove).

A2. Angular and velocity distribution of reactive molecules scattered from platinum surfaces (1982, H. Lin, S. T. Ceyer, M. Ascher).

D. Model of working hydrocarbon conversion catalysts. The role of carbonaceous deposits. (1982, F. Zaera, S. M. Davis).

D. Rhenium catalyzed ammonia synthesis (1982, N. D. Spencer).

D.   Effects of coadsorbed potassium on molecules—on platinum surface (1983, E. L. Garfunkel, J. C. Frost, J. E. Crowell).

D.   Catalytic hydrodesulfurization of thiophene on molybdenum crystal surfaces (1984, A. J. Gellman, M. E. Bussell, W. T. Tysoe).

A5.   High resolution electron energy loss spectroscopy of hydrocarbon molecules adsorbed on rhodium and platinum (1984, B. E. Bent and B. E. Koel).

I.   Catalyzed photodissociation of water (1988, G. H. Vurens, C. Sanchez).

D.   Effects of potassium and alumina on iron crystal surfaces in ammonia synthesis (1988, S. Bare, D. R. Strongin).

A6.   High pressure scanning tunneling microscopy of adsorbed atoms and molecules on metals (1988—B. Marchon, M. Salmeron, D. F. Ogletree, B. J. McIntyre, P. Sautet, J. C. Dunphy, K. B. Rider, A. Yoon, David Tang, M. Montano, D. Butcher, F. Tao)

A1.   Coadsorption induced ordering on metal surfaces (1988, C. M. Mate, G. S. Blackman, B. E. Bent, M. A. Van Hove and H. Ohtani)

I.   Titanium oxide overlayers on rhodium: adsorption and catalytic reactions. (1988, K. J. Williams, M. E. Levine, A. B. Boffa)

A1.   Adsorbate-induced restructuring of metal surfaces (1989, M. A. Van Hove, B. E. Bent).

A1.   Real-space multiple scattering theory of LEED (1990, M. A. van Hove, P. J. Rous).

I.   Silicon carbide and carbon films: Plasma deposition. (1991, G. J. Vandentop, M. P. Delplancke, J. M. Powers).

A1.   Disordered adsorption structure of benzene on platinum by diffuse LEED (1991, G. Held, R. Q. Hwang, M. A. van Hove, U. Starke, N. Materer).

A1.   Iron oxide overlayers on platinum (1992, G. Vurens, W. Weiss, A. Barbieri, M. A. van Hove).

A1.   LEED study of ice on platinum (1993, U. Starke, N. Materer, Ch. Minot, M. A. Van Hove).

D.   Catalyzed hydrocarbon reactions over platinum modified by rhenium, sulfur, potassium and tin (1994, F. H. Ribeiro, A. Bonevardi, T. Fujikawa).

A7.   Sum frequency generation surface vibrational spectroscopy studies of adsorbates and catalytic reactions on metals (1995— P. Cremer, C. Stanners, X. Su, J. Lahtinen, G. Rupprechter, K. McCrea, K. Y. Kung, M. Yang, A. L. Marsh.

I.   Surface science and catalysis of model Ziegler-Natta catalyst—$MgCl_2$ on gold with titanium chloride (1995, E. Magni, S. H. Kim, C. R. Tewell).

C.   Electron beam lithography of metal nanoparticle deposition (1996, P. W. Jacobs, J. Jensen, D. Gracias, Ji Zhu, M. X. Yang, A. Eppler, A. Avoyan, J. Grunes,)

A7, F. Sum frequency generation studies of polymer surfaces (1997, D. Zhang, R. S. Ward, D. Gracias, A. S. Eppler, Z. Chen, S. H. Kim, A. Opdahl, T. S. Koffas, E. Amitay-Sadovsky, S. Kwenskin, K. C. Chou, O. Mermut, R. L. York, A. M. Contreras, M. Koebel, G. J. Holinga)

A7.   Sum frequency generation studies of ice (1998—X. Su, C. Minot, M. A. Van Hove).

A1.   LEED surface crystallography of alkali halide thin films on platinum crystal surfaces (1999—J. G. Roberts, S. Hoffer, M. A. Van Hove)

I.   Sum frequency generation studies of platinum electrode surfaces with external potential during adsorbtion (1999—S. Baldelli, B. Mailhot).

C.   Colloid synthesis of metal nanoparticles (2002—Z. Konya, I. Kiricsi, R. M. Rioux, H. Song, J. D. Hoefelmeyer, K. Niesz, J. Grunes, M. Grass, Y. Bordoko, Y. Zhang, S. M. Humphrey, J. Kuhn, F. Tao, W. Huang, M. Koebel, S.H. Joo, K. An).

I.   Supersonic nozzle chemistry of hydrocarbons (2000—L. Romm).

A7.   Sum frequency generation studies of catalytic reaction intermediates (2004—K. C. Chou, S. Westerberg, M. Yang, A. Opdahl, K. M. Bratlie, C. J. Kliewer, L. Flores, K. R. McCrea).

G.   Hot electron flow during catalytic reaction (2005—X. Z. Ji, J. Y. Park, J. R. Renzas, Yawen Zhang, Antoine Hervier, Robert Baker, Hyungtak Seo, Griffin Kennedy).

E.   Nanoparticle catalysis (2007—Katie Bratlie, R. M. Rioux, H. Lee, K. Niesz,, Chris Kliewer, M. Bieri, C.-K. Tsung, J. Kuhn, W. Huang, J. R. Renzas, Y. Zhang, V. Pushkarev, S. Alayoglu).

B3, 4. Synchrotron studies ambient pressure XPS, EXAFS, and NEXAFS (2008—Y. Lee, M. Grass, H. Blum, F. Tao, V. Pushkarev, S.K. Beaumont, S. Alayoglu, F. Zhang, Z. Zhu).

A7,E.   Sum frequency generation studies of nanoparticles (2009—C. Aliaga, F. Tao, C.-K. Tsung, Y. Borodko, S. Alayoglu, J. M.Krier).

F.   Sum frequency generation studies of biointerfaces (2009, J. Kim, R. L. York, G. J. Holinga, R.M. Onorato, C. M. Thompson).

H.   Bridging heterogeneous and homogeneous catalysis (2010—Cole Witham, W. Huang, C.-K. Tsung, E. Gross, Y. Lee).

C.   Bimetallic nanoparticles (J. R. Renzas, W. Huang, Y. Zhang, M. Grass, S. Alayoglu, J. Park, S.-H. Joo).

C.   Tandem catalysis (2011—C.-K. Tsung).

AWARD CITATIONS

I have been lucky to attract so many wonderful and gifted students and research associates. I gratefully share the accolades and words of praise in the following citations from national and international awards with my hardworking research group.

<u>ACS Colloid and Surface Chemistry Award (1981)</u>: *For his imaginative and stimulating work on the structure of catalytic surfaces and adsorbed layers as well as the study of reactions proceeding on the surfaces.*

<u>The Peter Debye Award in Physical Chemistry (1989)</u>: *Gabor A. Somorjai. For his pioneering advocacy in placing heterogeneous catalysis on*

a fundamental molecular basis and for his career of creative research that has gone a long way towards establishing this goal.

<u>MRS Van Hippel Award (1997)</u>: *For his extraordinary multidisciplinary contributions to the atomic level understanding of materials surfaces and surface processes with technological importance in heterogeneous catalysis, corrosion and tribology*

<u>Wolf Prize in Chemistry (1998)</u>: For his outstanding contributions to the field of surface science in general and for the elucidation of fundamental mechanisms of heterogeneous catalytic reactions at single crystal surfaces in particular

<u>Linus Pauling Medal (2000)</u>: *For outstanding achievement in Chemistry*

<u>ACS Award for Creative Research in Homogeneous and Heterogeneous Catalysis (2000)</u>: *For his development of model catalysts and the molecular surface science of heterogeneous catalysis.*

<u>National Medal of Science (2002)</u>: *Gabor A. Somorjai. For molecular studies of surfaces through the use of single crystals and the development of new techniques that served as foundations of new surface technologies including heterogeneous catalysis.*

<u>Cotton Medal (2003)</u>: *For excellence in research in Chemistry*

<u>APS Irving Langmuir Prize (2007)</u>: *Gabor A. Somorjai. For his pioneering research in surface chemistry and delineation of catalytic mechanisms.*

<u>Priestley Medal (2008)</u>: *Gabor A. Somorjai. For his extraordinarily creative and original contributions to surface science and catalysis*

Honda Prize (2011): *For his pioneering contributions to surface chemistry that established the foundation of today's sophisticated catalysis*

The ENI Prize (2011): *Gabor Somorjai led fundamental researches on catalysis focusing above all on the preparation and characterization of supported metal catalysts. He carried out outstanding studies concerning the relationship between homogeneous and heterogeneous catalysis, in particular on the reactions of cyclization and hydroformylation. Professor Somorjai developed new catalysts based on nanostructured supported metal systems able to provide excellent performances and stabilities at high pressures.*

The BBVA Foundation Frontiers of Knowledge Award in Basic Sciences (2011): *Professor Gabor A. Somorjai for his pioneering experimental and conceptual contributions to the understanding of surface chemistry and catalysis at the molecular level with enormous repercussions in every-day and economic life.*

National Academy of Sciences Award in Chemical Sciences (2013). *For innovative research in the chemical sciences that contribute to a better understanding of the natural sciences and to the benefit of humanity.*

Index

bold denotes photo; *i* denotes illustration

energy conversion, 122, 176, 177, 194

Energy Reorganization Act (1974), 60

energy research, 60

Energy Research and Development Administration (ERDA), 60, 121

energy storage, 177, 194

energy technologies, 120–21

Engelhard and Degussa, 130

"engine knocking," 130

engineered knees and hips, 153

ENI Foundation (Italy), 179

environmental benefits, 120, 194

environmental clean-up technology, 89

enzymes, 176, 176i

ERDA (Energy Research and Development Administration), 60, 121

Ertl, Barbara, 189

Ertl, Gerhard, 110, 158, 187, 189

ESCA (Electron Spectroscopy for Chemical Analysis), 104

ETH (Swiss Federal Institute of Technology)
    author offered research director position at, 185–86
    honorary degree from, 179
    visit to (1987), 147

ethylene hydrogenation, 135, 165, 168

ethylidyne on Pt (111), 125i

Europe
    sabbatical in (1969-1970), 107–11
    visit to (1974), 111–12

experimental physical chemistry, 64

experimentation, approaches to, 64, 123, 139, 163

Exxon (ExxonMobile), 66

F

federal support for scientific research (US), 58–61, 196–97

Ferrer, Salvador, 184

FHI (Fritz Haber Institute), 148, 158, 187

field emission microscopy, 99, 107

field ion microscopy, 99, 107

Fischer-Tropsch synthesis, 122

flaring out process, 80

France, Anatole, 18

Free Speech Movement (Berkeley), 86–88

Free University of Brussels, 148

Freund, Hajo, 148, 187

friction, 108, 150, 151, 152, 194

Fritz Haber Institute (FHI), 148, 158, 187

Frontiers of Hydrocarbon Prize (2011), 179

Materials Research Society, Von
  Hippel Award (1997), 158
materials science, 157, 162, 177,
  194
Mauthausen, Austria, 10
Max Planck Society, 148
McCarthy, Joseph, 61
McCarthyism, 61
McMillan, Edwin M., 53
mean free path, 79
Meitner, Lise, 104
Metallurgical Laboratory,
  University of Chicago, 50
metal-surface-bound molecules,
  135
microelectronics, xiii, 106, 162,
  164
microprocessors, 194
Miller indices, 117, 118
"minimalistic" approach to
  experimentation, 64, 123
Minot, Christian, 112, 182
Minta Gymnasium, 22
modern surface science, 100, 122,
  151, 188
molecular adsorption, 125*i*
molecular beam surface scattering
  experiments, 112–16, 113*i*, 120,
  155, 156
molecular scale surface reaction
  phenomena, 140
molecular-scale surface
  information, 79

monodisperse nanoparticles, 168
monographs, 109–10
Montecatini
  author as consultant to, 139
  Paolo Galli at, 183
Müller, Erwin, 107
multicomponent nanoparticles,
  168

N

Nagy, Ferenc, 20
Nagy, Imre, 37, 38
Nano Letters, 102
nanocatalysis, 157, 161–78
nanomaterials synthesis, 156
nanometer-scale science, 161
nanoparticle surfaces, 163*i*
nanoscale patterning, 163
nanotechnology, 162
National Academies, 131, 132
National Academy of Engineering,
  132
National Academy of Sciences
  (NAS), author inducted into,
  131, 132
National Academy of Sciences
  (NAS), history of, 131–32
National Aeronautics and Space
  Administration (NASA), 59
National Institutes of Health, 61,
  132, 197
National Medal of Science

Wolf Prize in Chemistry (1998),
    158
World University Service, 41, 44

X

Xide, Xie, 190
X-ray crystallography, 63, 65
X-ray photoelectron spectroscopy
    (XPS), 88, 103, 104, 120, 141,
    154, 173

Y

Yang, Peidong, 177
Ynduráin, Felix, 184

Z

Zweig, Stefan, 18